PENGUIN BOOKS

FUNDAMENTALS

Frank Wilczek won the Nobel Prize in Physics in 2004 for work he did as a graduate student. He was among the earliest MacArthur fellows, and has won many awards both for his scientific work and his writing. He is the author of *A Beautiful Question*, *The Lightness of Being*, *Fantastic Realities*, *Longing for the Harmonies*, and hundreds of articles in leading scientific journals. His "Wilczek's Universe" column appears regularly in the *Wall Street Journal*. Wilczek is the Herman Feshbach Professor of Physics at the Massachusetts Institute of Technology, founding director of the T. D. Lee Institute and chief scientist at the Wilczek Quantum Center in Shanghai, China, and a distinguished professor at Arizona State University and Stockholm University.

Praise for *Fundamentals*

"*Fundamentals* might be the perfect book for the winter of this plague year. . . . Wilczek writes with breathtaking economy and clarity, and his pleasure in his subject is palpable."

—*The New York Times Book Review*

"For those with more scientific yearnings, and who regret not taking a few courses in college to learn about the physical world, theoretical physicist Frank Wilczek offers a way to catch up. . . . With his clear and joyful voice,

Wilczek succeeds very well, and for good reason. . . . There is no calculus required; this is not Physics 101. Instead, Wilczek talks about modern physics and cosmology from a more broad-brush and philosophical perspective, often linking their findings to the real world—how they affect us. In this age of rising skepticism, he wants his readers—whom he imagines to be lawyers, doctors, artists, parents, or simply curious people—to be 'born again, in the way of science.'"

—*The Washington Post*

"A lucid and riveting narrative of the fundamentals—what Wilczek calls 'the central messages of modern physics,' which are not just facts about how the world works but also 'the style of thought that allowed us to discover them.'"

—*Scientific American*

"*Fundamentals* is an engaging account of the history of humankind's understanding of reality, told by one of the key contributors to recent parts of that story. Wilczek's grasp on the physics he relates is comprehensive and authoritative; he conveys technicalities with a rare combination of accuracy and accessibility. . . . Wilczek provides an exceptionally clear guide to the state of physical knowledge in the early twenty-first century, much in the spirit of the sort of explanation that the ancient Greeks desired."

—*Science News*

"[Wilczek] turns out to be a true visionary."

—*The Times* (London)

"A breathtaking feat . . . The narrative is a mind-bender of the first order—in the best way possible—but what makes it so engrossing is that the author does far more than just present the facts and speculations, however fascinating; on every page, readers will glean his exhilaration and joy in discovery. . . . Another winner from Wilczek, who invites us to be born again into a richer, deeper understanding of the world."

—*Kirkus Reviews* (starred review)

"[A] delightful book . . . Frank Wilczek is that rare creature: a first-class scientist who is also an extremely talented communicator. . . . Wilczek constantly finds fresh ways to present such ideas, so that you emerge with new insight into what they mean. . . . *Fundamentals* is, then, not only an exceptional piece of science communication but also a deeply humanistic book. It celebrates what we know without pretending that is more than it is: 'The world is complex beyond our ability to grasp, and rich in mysteries, but we know a lot, and are learning more. Humility is in order, but so is self-respect.'"

—*Physics World*

"This is an exuberant, gorgeously crafted, and intellectually thrilling book, written by one of our greatest living scientists yet hospitable to all. To be reminded that time and space, mystery and order, are so much stranger and more generous than we can comprehend—this is a gift to public life and moral imagination in a young century where what is visible and tangible feels chaotic and constricting. . . . What began as an exposition, as Wilczek writes, 'grew into a contemplation.' The result is a profoundly enriched understanding, accessible to the religious and non-religious alike, of what it means to be human—and what we might be pointing at when we use the word God."

—Krista Tippett, host of *On Being* and author of *Becoming Wise*

"This is a book about deep ideas, not passing fancies. It will teach you profound principles, not dry lists of facts. It's a rare treat indeed to get a glimpse into the mind of one of the world's leading physicists, presented in an engaging style that will be enjoyed by anyone at all."

—Sean Carroll, author of *Something Deeply Hidden*

"Frank Wilczek is not only one of the world's greatest physicists; he's also one of its greatest explainers. *Fundamentals* is lucid, beautiful, and revelatory."

—Steven Strogatz, professor of mathematics at Cornell University and author of *Infinite Powers*

"If you were to go back just two hundred years and tell people what we knew, from the origins of the universe to the molecular basis of life, and how weird and unintuitive nature is at the atomic scale, they would think we were crazy. But if you showed them what we have created with that knowledge, they would think we were magicians. In this engaging and highly accessible book, Frank Wilczek shows how the vast edifice that is modern science was constructed with only a few ingredients and assumptions, but depended crucially on a way of thinking—about the nature of evidence and how it applied to the world around us. Anyone interested in the underlying basis of the complexity of today's science will enjoy this book."

—Venki Ramakrishnan, winner of the Nobel Prize in Chemistry and author of *Gene Machine*

"Few living scientists have accomplished more than Frank Wilczek in helping unveil that deeper layer of existence. With poetry and fervor, Wilczek takes us on a breathtaking journey to the frontiers of physics, and reminds us of just how privileged we human beings are to glimpse the foundations of reality."

—Paul Davies, Regents Professor at Arizona State University and author of *The Demon in the Machine*

"This book is a gem. As a schoolboy, I devoured popular science books by Gamow and Asimov, but kept wondering if anyone could succeed as admirably with modern theoretical physics, especially particle physics and cosmology with their counterintutive concepts and specialized jargon. The answer is the book that you now hold in your hands. The author manages to inspire us, transporting us into the wonderland of modern physics without ever compromising on facts."

—V. S. Ramachandran, author of *The Tell-Tale Brain*

"*Fundamentals* reflects Frank Wilczek's extraordinary intellectual range as one of the most venerated, original, and daring physicists of our time. Frank's admiration for nature and mathematical beauty infuses his ruminations on 'ten keys to reality,' his curated collection of the most profound physical touchstones. Through the ten chapters, the reader has the privilege of following this great mind from discoveries into the realm of speculation for an exceptional opportunity to revel in the mysteries of the universe."

—Janna Levin, author of *Black Hole Survival Guide*

Fundamentals

TEN KEYS TO REALITY

Frank Wilczek

PENGUIN BOOKS

PENGUIN BOOKS
An imprint of Penguin Random House LLC
penguinrandomhouse.com

First published in the United States of America by Penguin Press,
an imprint of Penguin Random House LLC, 2021
Published in Penguin Books 2022

ISBN 9780735223905 (paperback)

THE LIBRARY OF CONGRESS HAS CATALOGED THE HARDCOVER EDITION AS FOLLOWS:
Names: Wilczek, Frank, author.
Title: Fundamentals : ten keys to reality / Frank Wilczek.
Description: New York : Penguin Press, 2021. | Includes index.
Identifiers: LCCN 2020020086 (print) | LCCN 2020020087 (ebook) |
ISBN 9780735223790 (hardcover) | ISBN 9780735223899 (ebook)
Subjects: LCSH: Reality.
Classification: LCC QC6.4.R42 W55 2021 (print) | LCC QC6.4.R42 (ebook) |
DDC 530.01—dc23
LC record available at https://lccn.loc.gov/2020020086
LC ebook record available at https://lccn.loc.gov/2020020087

Printed in the United States of America
1st Printing

Designed by Amanda Dewey

For Betsy:

REVELATIONS

Orchestrated multitudes spin out

The textured patterns which make our lives.

Birth, learning, love, and unsought age—

Gifts we did not earn, limits we did not grant.

Space grows in silence, past our grasp.

Celestial bodies, thinly sprinkled there

Broadcast in obedience to ideal laws.

They do not speak the language sung at cradles.

Time is change, impartially enforced.

In ancient things we see its awesome scope

While tiny, perfect clocks attest its vigor.

Time long predates us, and will long outlive us.

As in my mind I make my world anew

The cherished, closest thing is ever you.

CONTENTS

PREFACE: BORN AGAIN

I

This is a book about fundamental lessons we can learn from the study of the physical world. I've met many people who are curious about the physical world and eager to learn what modern physics says about it. They might be lawyers, doctors, artists, students, teachers, parents, or simply curious people. They have intelligence, but not knowledge. Here I've tried to convey the central messages of modern physics as simply as possible, while not compromising accuracy. I've kept my curious friends and their questions constantly in mind while writing the book.

To me, those fundamental lessons include much more than bare facts about how the physical world works. Those facts are both powerful and strangely beautiful, to be sure. But the style of thought that allowed us to discover them is a great achievement, too. And it's important to consider what those fundamentals suggest about how we humans fit into the big picture.

II

I've selected ten broad principles as my fundamentals. Each forms the theme of one chapter. In the body of each chapter, I explain and document that chapter's theme from different perspectives, and then make some informed guesses about its future development. Those informed guesses were fun to create, and I hope they're exciting to read. They are meant to convey another fundamental message: that our understanding of the physical world is still growing and changing. It is a living thing.

I've been careful to separate speculations from facts and, for the facts, to indicate the nature of the observations and experiments that establish them. For perhaps the most fundamental message of all is that we *do* understand many aspects of the physical world very deeply. As Albert Einstein put it, "The fact that [the universe] is comprehensible is a miracle." That, too, was a hard-won discovery.

Precisely because it is so surprising, the comprehensibility of the physical universe must be demonstrated, not assumed. The most convincing proof is that our understanding, though incomplete, has let us accomplish great and amazing things.

In my research, I try to fill gaps in our understanding and to design new experiments to push the frontiers of possibility. It's been a joy for me, in writing this book, to step back and reflect, wonderstruck, on some highlights of what generations of scientists and engineers, cooperating across time and space, have already accomplished.

III

Fundamentals is meant, as well, to offer an alternative to traditional religious fundamentalism. It takes up some of the same basic questions, but addresses them by consulting physical reality, rather than texts or traditions.

Many of my scientific heroes—Galileo Galilei, Johannes Kepler, Isaac Newton, Michael Faraday, James Clerk Maxwell—were devout Christians. (In this they were representative of their times and surroundings.) They thought that they could approach and honor God by studying His work. Einstein, though he was not religious in a conventional sense, had a similar attitude. He often referred to God (or "the Old One"), as he did in one of his most famous quotations: "Subtle is the lord, but malicious he is not."

The spirit of their enterprise, and mine here, transcends specific dogmas, whether religious or antireligious. I like to state it this way: In studying how the world works, we are studying how God works, and thereby *learning what God is*. In that spirit, we can interpret the search for knowledge as a form of worship, and our discoveries as revelations.

IV

Writing this book changed my perception of the world. *Fundamentals* began as an exposition but grew into a contemplation. As I reflected on the material, two overarching themes emerged unexpectedly. Their clarity and depth have astonished me.

The first of those themes is abundance. The world is large. Of course, a good look at the sky on a clear night is enough to show you that there's lots of space "out there." When, after more careful study, we put numbers to that size, our minds are properly boggled. But the largeness of space is only one aspect of Nature's abundance, and it is not the one most central to human experience.

For one thing, as Richard Feynman put it, "there's plenty of room at the bottom." Each of our human bodies contains far more atoms than there are stars in the visible universe, and our brains contain about as many neurons as there are stars in our galaxy. The universe within is a worthy complement to the universe beyond.

As for space, so also for time. Cosmic time is abundant. The quantity of time reaching back to the big bang dwarfs a human lifetime. And yet, as we'll discuss, a full human lifetime contains far more moments of consciousness than universal history contains human lifespans. We are gifted with an abundance of inner time.

The physical world is abundant, as well, in hitherto untapped resources for creation and perception. Science reveals that the nearby world contains, in known and accessible forms, far more energy and usable material than humans presently exploit. This realization empowers us and should whet our ambitions.

Our unaided perception brings in only a few slivers of the reality that scientific investigation reveals. Consider, for example, vision. Our sense of vision is our widest and most important portal to the external world. But it leaves so much unseen! Telescopes and microscopes reveal vast treasure troves of information, encoded in light, that ordinarily come to our eyes unrecognized. Moreover, our vision is limited to one octave—the span of visible light—from an infinite keyboard of electromagnetic radiation, which runs from radio waves to microwaves to infrared on one side, and from ultraviolet to x-rays and gamma rays on the other. And even within our one octave, our color vision is blurry. While our senses fail to perceive many aspects of reality, our minds allow us to transcend our natural limits. It is a great, continuing adventure to widen the doors of perception.

V

The second theme is that to appreciate the physical universe properly one must be "born again."

As I was fleshing out the text of this book, my grandson Luke was born. During the drafting, I got to observe the first few months of his life. I saw how he studied his own hands, wide-eyed, and began to realize that he controlled them. I saw the joy with which he learned to reach out and grasp objects in the external world. I watched him experiment with objects, dropping them and searching for them, and repeating himself (and repeating himself . . .), as if not quite certain of the result, but laughing in joy when he found them.

In these and many other ways, I could see that Luke was constructing a model of the world. He approached it with insatiable curiosity and few preconceptions. By interacting with the world, he learned the things that nearly all human adults take for granted, such as that the world divides into self and not-self, that thoughts can control movements of self but not of not-self, and that we can look at bodies without changing their properties.

Babies are like little scientists, making experiments and drawing conclusions. But the experiments they do are, by the standards of modern science, quite crude. Babies work without telescopes, microscopes, spectroscopes, magnetometers, particle accelerators, atomic clocks, or any other of the instruments we use to construct our truest, most accurate world-models. Their experience is limited to a small range of temperatures; they are immersed in an atmosphere with a very special chemical composition and pressure; Earth's gravity pulls them (and everything in their environment) down, while Earth's surface supports them . . . and so forth.

Babies construct a world-model that accounts for what they experience *within the bounds of their perception and environment.* For practical purposes, that's the right plan. To cope with the everyday world, it is efficient, and reasonable, when we are children, to take lessons from the everyday world.

But modern science reveals a physical world very different from the model we construct as babies. If we once again open ourselves up to the world, curious and without preconceptions—if we allow ourselves to be born again—we come to understand the world differently.

Some things, we must learn. The world is built from a few basic building blocks, which follow strict but strange and unfamiliar rules.

Some things, we must unlearn.

Quantum mechanics reveals that you cannot observe something without changing it, after all. Each person receives unique messages from the external world. Imagine that you and a friend sit together in a very dark room, observing a dim light. Make the light very, very dim, say, by covering it with layers of cloth. Eventually, both you and your friend will see only intermittent flashes. But you will see flashes at different times. The light has broken up into individual quanta, and quanta cannot be shared. At this fundamental level, we experience separate worlds.

Psychophysics reveals that consciousness does not direct most actions, but instead processes reports of them, from unconscious units that do the work. Using a technique known as transcranial magnetic stimulation (TMS), it is possible to stimulate the left or right brain motor centers in a subject's brain, at the experimenter's discretion. A properly sculpted TMS signal to the right motor center will cause a twitch of the left wrist, while a properly sculpted TMS signal to the left motor center will cause a twitch of the right wrist. Alvaro Pascual-Leone used this technique ingeniously in a simple experiment that has profound implications. He asked subjects, upon receiving a cue, to decide whether they wanted to twitch their right or their left wrist. Then they were instructed to act out their intention upon receiving an additional cue. The subjects were in a brain scanner, so the experimenter could watch their motor areas preparing the twitch. If they had decided to twitch their right wrist, their left motor area was active; if they decided to twitch their left wrist, their right motor area was active. It was possible, in this way, to predict what choice had been made before any motion occurred.

Now comes a revealing twist. Occasionally Pascual-Leone would apply a TMS signal to contradict (and, it turns out, override) the subject's choice. The subject's twitch would then be the one that TMS imposed, rather than the one he or she originally chose. The remarkable thing is how the subjects explained what had happened. They did *not* report that some external force had possessed them. Rather, they said, "I changed my mind."

Detailed study of matter reveals that our body and our brain—the physical platform of our "self"—is, against all intuition, built from the same stuff as "not-self," and appears to be continuous with it.

In our rush to make sense of things, as infants, we learn to misunderstand the world, and ourselves. There's a lot to unlearn, as well as a lot to learn, on the voyage to deep understanding.

VI

The process of being born again can be disorienting. But, like a roller-coaster ride, it can also be exhilarating. And it brings this gift: To those who are born again, in the way of science, the world comes to seem fresh, lucid, and wonderfully abundant. They come to live out William Blake's vision:

> To see a World in a Grain of Sand
> And a Heaven in a Wild Flower
> Hold Infinity in the palm of your hand
> And Eternity in an hour

INTRODUCTION

I

The universe is a strange place.

To newborn infants, the world presents a jumble of bewildering impressions. In sorting it out, a baby soon learns to distinguish between messages that originate from an internal world and those that originate from an external world. The internal world contains both feelings, such as hunger, pain, well-being, and drowsiness, and the netherworld of dreams. Within it, too, are private thoughts, such as those that direct her gaze, her grasp, and, soon, her speech.

The external world is an elaborate intellectual construction. Our baby devotes much of her time to making it. She learns to recognize stable patterns in her perception that, unlike her own body, do not respond reliably to her thoughts. She organizes those patterns into objects. She learns that those objects behave in somewhat predictable ways.

Eventually our baby, now a child, comes to recognize some of the objects as beings similar to herself, beings with whom she can communicate. After exchanging information with those beings, she becomes convinced that they, too, experience internal and external worlds and, remarkably, that all of them share many objects in common, and that those objects obey the same rules.

II

Understanding how to control the common external world—in other words, the physical world—is, of course, a vital practical problem, with many aspects. For example, to thrive in a hunter-gatherer society, our child would have to learn where to find water; which plants and animals are good to eat, and how to find, raise, or hunt them; how to prepare and cook food, and many other facts and skills.

In more complex societies, other challenges arise, such as how to make specialized tools, how to build lasting structures, and how to keep track of time. Successful solutions to the problems posed by the physical world get discovered, shared, and accumulated over generations. They become, for each society, its "technology."

Nonscientific societies often develop rich and complex technologies. Some of those technologies enabled—and still do enable—people to thrive in difficult environments, such as the Arctic or the Kalahari Desert. Others supported the construction of great cities and impressive monuments, such as the Egyptian and Mesoamerican pyramids.

Still, throughout most of human history, prior to the emergence of the scientific method, the development of technologies was haphazard. Successful techniques were discovered more or less by accident. Once stumbled upon, they were transmitted in the form of very specific procedures, rituals, and traditions. They did not form a logical system, nor was there a systematic effort to improve them.

Technologies based on "rules of thumb" allowed people to survive, reproduce, and, often, to enjoy some leisure and achieve satisfying lives. For most people, in most cultures, over most of history, that was enough. People had no way to know what they were missing, or that what they were missing might be important to them.

But now we know that they were missing a lot. This figure, which shows the development of human productivity with time, speaks for itself, and it speaks volumes.

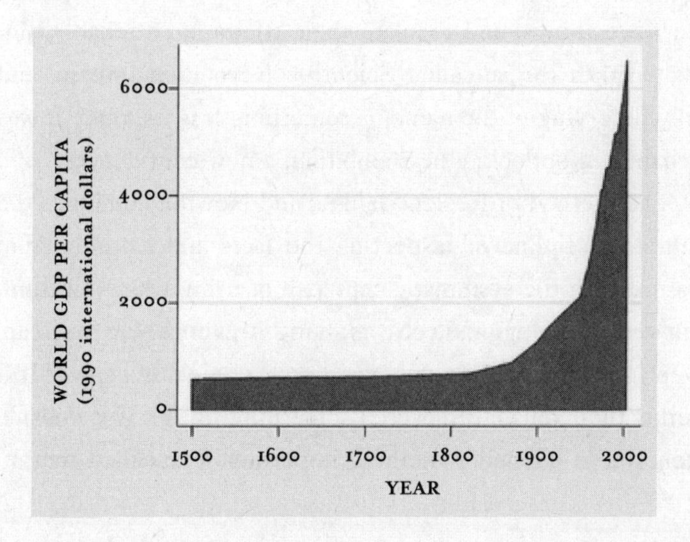

III

The modern approach to understanding the world emerged in Europe in the seventeenth century. There were partial anticipations earlier, and elsewhere. But the constellation of breakthroughs known as the Scientific Revolution provided inspiring examples of what could be achieved by human minds creatively engaged with the physical world, and the methods and attitudes that led to those breakthroughs gave clear models for future exploration. With that impetus, science as we know it began. It has never looked back.

The seventeenth century saw dramatic theoretical and technological progress on many fronts, including in the design of mechanical machines and ships, of optical instruments (including, notably, microscopes and telescopes), of clocks, and of calendars. As a direct result, people could wield more power, see more things, and regulate their affairs more reliably. But what makes the so-called Scientific Revolution unique, and fully deserving of the name, is something less tangible. It was a change in outlook: a new ambition, a new confidence.

The method of Kepler, Galileo, and Newton combines the humble discipline of respecting the facts and learning from Nature with the systematic chutzpah of using what you think you've learned aggressively, applying it everywhere you can, even in situations that go beyond your original evidence. If it works, then you've discovered something useful; if it doesn't, then you've learned something important. I've called that attitude Radical Conservatism, and to me it's the essential innovation of the Scientific Revolution.

Radical Conservatism is conservative because it asks us to learn from Nature and to respect facts—key aspects of what is called the scientific method. But it is radical, too, because it pushes what you've learned for all it's worth. This is no less essential to how science actually works. It provides science with its cutting edge.

IV

This new outlook was inspired, above all, by developments in a subject that even in the seventeenth century was already ancient and well developed: celestial mechanics, the description of how objects in the sky appear to move.

Since long before the beginning of written history, people have recognized such regularities as the alternation of night and day, the cycle of seasons, the phases of the Moon, and the orderly procession of stars. With the rise of agriculture, it became crucial to keep track of seasons, in order to plant and harvest at the most appropriate times. Another powerful, if misguided, motivation for accurate observations was the belief that human life was directly connected to cosmic rhythms: astrology. In any case, for a mixture of reasons—including simple curiosity—people studied the sky carefully.

It emerged that the vast majority of stars move in a reasonably simple, predictable way. Today, we interpret their apparent motion as resulting from Earth rotating around its axis. The "fixed stars" are so far away that relatively small changes in their distance, whether due to their own proper motion or to the motion of Earth around the Sun, are invisible to the naked eye. But a few exceptional objects—the Sun, the Moon, and a few "wanderers," including the naked-eye planets Mercury, Venus, Mars, Jupiter, and Saturn—do not follow that pattern.

Ancient astronomers, over many generations, recorded the positions of those special objects, and eventually learned how to predict their changes with fair accuracy. That task required calculations in geometry and trigonometry, following complicated, but perfectly definite, recipes. Ptolemy (c. 100–170) brought this material together in a mathematical text that became known as *Almagest*. (*Magest* is a Greek superlative meaning "greatest." It has the same root as "majestic." *Al* is simply Arabic for "the.")

Ptolemy's synthesis was a magnificent achievement, but it had two shortcomings. One was its complexity and, related to this, its ugliness. In particular, the recipes it used to calculate planetary motions brought in many numbers that were determined purely by fitting the calculations to observations, without deeper guiding principles connecting them. Copernicus (1473–1543) noticed that the values of some of those numbers were related to one another in surprisingly simple ways. These otherwise mysterious, "coincidental" relationships could be explained geometrically, if one assumed that Earth together with Venus, Mars, Jupiter, and Saturn all revolve around the Sun as center (and the Moon further revolves around Earth).

The second shortcoming of Ptolemy's synthesis is more

straightforward: It simply isn't accurate. Tycho Brahe (1546–1601), in an anticipation of today's "Big Science," designed elaborate instruments and spent a lot of money building an observatory that enabled much more precise observations of planetary positions. The new observations showed unmistakable deviations from Ptolemy's predictions.

Johannes Kepler (1571–1630) set out to make a geometric model of planetary motion that was both simple and accurate. He incorporated Copernicus's ideas and made other important technical changes to Ptolemy's model. Specifically, he allowed the planetary orbits around the Sun to deviate from simple circles, substituting ellipses, with the Sun at one focus. He also allowed the rate at which the planets orbit the Sun to vary with their distance from it, in such a way that they sweep out equal areas in equal times. After those reforms, the system was considerably simpler, and it also worked better.

Meanwhile, back on the surface of Earth, Galileo Galilei (1564–1642) made careful studies of simple forms of motion, such as the way balls roll down inclined planes and how pendulums oscillate. Those humble studies, putting numbers to positions and times, might seem pitifully inadequate to addressing big questions about how the world works. Certainly, to most of Galileo's academic contemporaries, concerned with grand questions of philosophy, they seemed trivial. But Galileo aspired to a different kind of understanding. He wanted to understand *something* precisely, rather than *everything* vaguely. He sought—and found—definite mathematical formulas that described his humble observations fully.

Isaac Newton (1643–1727) weaved together Kepler's geometry of planetary motion and Galileo's dynamical description of motion on Earth. He demonstrated that both Kepler's theory of planetary motions and Galileo's theory of special motions were best understood as special cases of general laws, laws that apply to all bodies everywhere and for all time. Newton's theory, which we now call classical mechanics, went from triumph to triumph, accounting for the tides on Earth, predicting the paths of comets, and empowering new feats of engineering.

Newton's work showed, by convincing example, that one could address grand questions by building up from a detailed understanding of simple cases. Newton called this method *analysis and synthesis.* It is the archetype of scientific Radical Conservatism.

Here is what Newton himself had to say about that method:

> As in mathematics, so in natural philosophy the investigation of difficult things by the method of analysis ought ever to precede the method of composition. This analysis consists of making experiments and observations, and in drawing general conclusions from them by induction. . . . By this way of analysis we may proceed from compounds to ingredients, and from motions to the forces producing them; and in general from effects to their causes, and from particular causes to more general ones till the argument end in the most general. This is the method of analysis: and the synthesis consists in assuming the causes discovered and established as principles, and by them explaining the phenomena preceding from them, and proving the explanations.

V

Before leaving Newton, it seems appropriate to add another quotation, which reflects his kinship with his predecessors Galileo and Kepler, and with all of us who follow in their footsteps:

> To explain all nature is too difficult a task for any one man or even for any one age. 'Tis much better to do a little with certainty & leave the rest for others that come after you.

A more recent quotation from John R. Pierce, a pioneer of modern information science, beautifully captures the contrast between the modern concept of scientific understanding and all other approaches:

> We require that our theories harmonize in detail with the very wide range of phenomena they seek to explain. And we insist that they provide us with useful guidance rather than with rationalizations.

As Pierce was acutely aware, this heightened standard comes at a painful price. It involves a loss of innocence. "We will never again understand nature as well as Greek philosophers did. . . . We know too much." That price, I think, is not too high. In any case, there's no going back.

I

What There Is

I

THERE'S PLENTY OF SPACE

PLENTY OUTSIDE *AND* PLENTY WITHIN

When we say that the something is big—be it the visible universe or a human brain—we have to ask: Compared with what? The natural point of reference is the scope of everyday human life. This is the context of our first world-models, which we construct as children. The scope of the physical world, as revealed by science, is something we discover when we allow ourselves to be born again.

By the standards of everyday life, the world "out there" is truly gigantic. That *outer plenty* is what we sense intuitively when, on a clear night, we look up at a starry sky. We feel, with no need for careful analysis, that the universe has distances vastly larger than our human bodies, and larger than any distance we are ever likely to travel. Scientific understanding not only supports but greatly expands that sense of vastness.

The world's scale can make people feel overwhelmed. The

French mathematician, physicist, and religious philosopher Blaise Pascal (1623–1662) felt that way, and it gnawed at him. He wrote that "the universe grasps me and swallows me up like a speck."

Sentiments like Pascal's—roughly, "I'm *very* small, I make no difference in the universe"—are a common theme in literature, philosophy, and theology. They appear in many prayers and psalms. Such sentiments are a natural reaction to the human condition of cosmic insignificance, when measured by size.

The good news is that raw size isn't everything. Our *inner plenty* is subtler, but at least equally profound. We come to see this when we consider things from the other end, bottom up. There's plenty of room at the bottom. In all the ways that really matter, we're abundantly large.

In grade school, we learn that the basic structural units of matter are atoms and molecules. In terms of those units, a human body is huge. The number of atoms in a single human body is roughly 10^{28}–1 followed by 28 zeros: 10,000,000,000, 000,000,000,000,000,000.

That is a number far beyond what we can visualize. We can name it—ten octillion—and, after some instruction and practice, we can learn to calculate with it. But it overwhelms ordinary intuition, which is built on everyday experience, when we never have occasion to count that high. Visualizing that many individual dots far exceeds the holding capacity of our brains.

The number of stars visible to unaided human vision, in clear air on a moonless night, is at best a few thousand. Ten octillion, on the other hand, the number of atoms within us, is about a million times the number of stars in the entire visible universe. In that very concrete sense, a universe dwells within us.

Walt Whitman (1819–1892), the big-spirited American poet, felt our inner largeness instinctively. In his "Song of Myself" he wrote, "I am large, I contain multitudes." Whitman's joyful celebration of abundance is just as grounded in objective facts as Pascal's cosmic envy, and it is much more relevant to our actual experience.

The world is large, but we are not small. It is truer to say that there's plenty of space, whether we scale up or down. One shouldn't envy the universe just because it's big. We're big, too. We're big enough, specifically, to contain the outer universe within our minds. Pascal himself took comfort from that insight, as he followed his lament that "the universe grasps me and swallows me up like a speck" with the consolation "but through thought I grasp it."

The abundance of space—both its outer and its inner plenty—is the main topic of this chapter. We'll look deeper into the hard facts, and then venture a bit beyond.

OUTER PLENTY: WHAT WE KNOW AND HOW WE KNOW IT

Prelude: Geometry and Reality

Scientific discussion of cosmic distances is built on the foundation of our understanding of physical space and how to measure distance: the science of geometry. Let us begin, therefore, with the relationship between geometry and reality.

Direct, everyday experience teaches us that objects can move from place to place without changing their properties. This leads us to the idea of "space" as a kind of receptacle, wherein nature deposits objects.

Practical applications in surveying, architecture, and navigation led people to measure distances and angles among nearby objects. Through such work, they discovered the regularities on display in Euclidean geometry.

As practical applications got more extensive and demanding, that framework held up impressively. So successful was Euclid's geometry, and so majestic is its logical structure, that critical tests of its validity as a description of physical reality were rarely undertaken. In the early nineteenth century, Carl Friedrich Gauss (1777–1855), one of the all-time great mathematicians, thought it was worth doing a reality check. He measured the angles in a triangle formed by three distant mountain stations in Germany and found that they added up to 180°, as Euclid predicts, within measurement uncertainties. Today's Global Positioning System (GPS) is based on Euclidean geometry. It performs millions of experiments like Gauss's every day, but on larger scales and with much greater precision. Let's take a quick look at its workings.

To get your position using GPS, you tap into broadcasts from a collection of artificial satellites high above Earth, which know where they are. (We'll come back to how *that* happens.) Currently there are more than thirty of these satellites strategically arranged around the globe. Their radio broadcasts don't translate into talk or music. Instead, they send out simple announcements of where they are, in a digital format tailored to computers. The announcements include time stamps, which specify when they were sent. Each satellite carries a superb atomic clock onboard. That clock ensures that the time stamps are accurate. Then:

1. Your GPS unit's receiver picks up some of the satellite signals. The unit, which also has access to signals from an extensive network of ground-based clocks, computes how long the different satellite signals took to arrive. Since those signals travel at a known speed—the speed of light—the transit times can be used to determine the satellites' distances.
2. Using those distances, the positions of the satellites, and Euclidean geometry, the computer determines a unique position of the source—that is, you—by triangulation.
3. The computer reports that result, and you learn where you are.

The fully implemented GPS adds many clever refinements, but that is its central idea. This system bears an uncanny resemblance to Albert Einstein's "thought design" of reference frames in his original paper on special relativity. In 1905, he anticipated using light beams and transit times to map out spatial positions. Einstein liked that idea because it uses a technique rooted in basic physics—the fixed speed of light—to map out space. Technology has a way of catching up with thought experiments.

As an exercise in visual imagination, you might try to convince yourself that your distances to four satellites—each in a known position—provide more than enough information to reconstruct your position.

(Here's a hint: Points at a given distance from a satellite lie on a sphere centered on that satellite. If we take two spheres, centered on two different satellites, they will either intersect in a circle or not at all. Since your location is somewhere in the intersection, they'd better intersect! Now consider how a third sphere, corresponding to a third satellite, intersects that circle. Generally, they will intersect at two points. Finally, the fourth satellite's sphere will single out one of those two points.)

Now let's return to the question of how the GPS's satellites know where *they* are. The technical details get complicated, but the underlying idea is simple: They start from known positions, and then they track their own motion. By putting those two pieces of information together, they calculate where they are.

In more detail: The satellites monitor their motion using onboard gyroscopes and accelerometers, like the ones in your iPhone. From the observed response of those instruments, the satellite's computer can read out the satellite's acceleration, using the physics of Newtonian mechanics. From that input, using calculus, it calculates how much the satellite has moved. Indeed, Newton invented calculus to solve problems like this.

If you review all the steps, you'll see that the engineers who designed the Global Positioning System built on many nonobvious assumptions. The system relies on the idea that the speed of light is constant. It uses atomic clocks, whose design and interpretation relies on advanced principles of quantum theory, to do accurate timing. It uses the tools of classical mechanics to calculate the position of the satellites it deploys. It also makes corrections for the effect, predicted by general relativity, that the rate of clocks depends slightly on their elevation above Earth. Clocks run slower near Earth's surface, where its gravitational field is stronger.

Since the Global Positioning System relies on so many other assumptions in addition to the validity of Euclidean geometry, we cannot claim that it provides a clean, pure test of that geometry. Indeed, the success of GPS is not a clean, pure test of any single principle. It is a complicated system, whose design relies on a tangled web of assumptions.

Any of those assumptions might be wrong or, to put it more diplomatically, only approximately true. If any of the assumptions that engineers assumed to be "approximately true" were significantly wrong, GPS would give inconsistent results. For instance, you might derive different positions from triangulating on different sets of satellites. Hard use can reveal hidden weaknesses.

Conversely, to the extent that GPS works, its success reinforces our confidence in *all* the underlying assumptions, including the assumption that Euclidean geometry describes, with good accuracy, the reality of spatial geometry on earthly scales. And so far, GPS has worked flawlessly.

More generally, science builds. The most advanced, adventurous experiments and technologies rely on tangled webs of underlying theories. When those adventurous applications hold up, they increase our confidence in their supporting webs. The fact that fundamental understanding forms a tangled, mutually reinforcing web of ideas will be a recurring theme in what follows.

Before concluding this prelude, I must add a qualification. When we come to consider space on gigantic cosmic scales, as we're about to do, or with exquisite precision, or in the vicinity of black holes, Euclidean geometry stops matching reality. Albert Einstein, in his special and general relativity theories (in 1905 and 1915, respectively), exposed its inadequacies theoretically and suggested how to get beyond them. Since then, his theoretical ideas have been confirmed in many experiments.

Einstein taught us, in special relativity, that when we claim to measure "distance" we must consider carefully what it is we're measuring and how we are measuring it. Real measurements take time, and things can move in time. What we can actually measure is separations between *events*. Events are located in both space and time. The geometry of events must be constructed within that larger framework: space-time, not just space. In general relativity, we learn further that the geometry of space-time can be warped by the influence of matter, or by waves of distortion that travel through it. (More on this in chapters 4 and 8.)

Within the more comprehensive frameworks of space-time and general relativity, Euclidean geometry serves as an approximation to more accurate theories. It is accurate enough for use in the many practical applications mentioned above. Surveyors, architects, and designers of space missions use Euclidean geometry because they can get away with it, and it eases their work. The more comprehensive theories, while more accurate, are much more complicated to use.

The fact that Euclidean geometry fails to provide a complete model of reality does not detract from its mathematical consistency nor invalidate its many successes. But it does confirm the wisdom of Gauss's fact-checking, radically conservative approach. The relationship between geometry and reality is a question for Nature to settle.

Surveying the Universe

Having taken the measure of nearby space, we can proceed to survey the cosmos. The primary tools in this endeavor are various kinds of telescopes. Besides the familiar telescopes that employ visible light, astronomers use telescopes that gather "light" from many other parts of the electromagnetic spectrum, including radio waves, microwaves, infrared, ultraviolet, x-rays, and gamma rays. There are also more exotic eyes on the sky, not based on electromagnetic radiation, notably including a very recent addition, gravitational wave detectors. I'll say more about those in later chapters.

Let me begin by highlighting the amazingly simple conclusions of this survey. Then I'll review how astronomers reached them. That is more complicated—though, given the context, still amazingly simple.

The most fundamental conclusion is that we find the same kind of material everywhere. Furthermore, we observe that the same laws apply everywhere.

Second, we find that matter is organized into a hierarchy of structures. Everywhere we look, we can recognize stars. They tend to cluster into galaxies, which commonly contain anywhere from a few million to billions of stars. Our own star, the Sun, has a retinue of planets and moons (and also comets, asteroids, the beautiful "rings" of Saturn, and other debris). Jupiter, the largest planet, has about one-thousandth the weight of the Sun, while Earth has about three-millionths the weight of the Sun. Despite their modest share of mass, planets and their moons should be especially dear to our hearts. We live on one, of course, and there are reasons to suspect that others might support new forms of life—if not in our solar system, then elsewhere. Astronomers have long suspected that other stars might have planets, but it is only recently that they've developed the technical strength to detect them. By now, hundreds of extrasolar planets have been discovered, and new discoveries keep flooding in.

Third, we find that all this stuff is sprinkled nearly uniformly throughout space. We find roughly the same density of galaxies in all directions, and at all distances.

Later we will refine and supplement these three fundamental conclusions, notably to bring in the big bang, "dark matter," and "dark energy." But their central message endures: One finds the same sorts of substances, organized in the same sorts of ways, spread uniformly over the visible universe, in vast abundance.

By now you may be wondering how astronomers arrive at such far-reaching conclusions. Let's have a closer look, while filling in concrete values of the sizes and distances.

It is not immediately obvious how to measure the distance to very distant objects. Obviously, you can't lay down rulers, stretch tape measures into the sky, or monitor time-stamped radio transmissions. Instead, astronomers use a bootstrap technique, called the *cosmic distance ladder*. Each rung of the ladder takes us to larger distances. We use our understanding at one rung to prepare us for the next.

We can start by surveying distances in the immediate neighborhood of Earth. Using similar techniques to GPS—that is, bouncing light (or radio signals) around, and measuring transit times—we can determine distances on Earth, and distances from Earth to other objects in the solar system. There are several other ways to do this, including some ingenious, though not very accurate, methods invented by the ancient Greeks. For present purposes, it is enough to note that all of these methods give consistent results.

Earth itself is a near-perfect sphere, whose radius is roughly 6,400 kilometers, or 4,000 miles. In this age of air travel, that is a readily comprehensible distance. It is roughly equal to the overland distance between New York and Stockholm, or slightly more than half the distance between New York and Shanghai.

There is another way of stating distance, which is beautifully adapted to astronomy and cosmology, and is widely used in those subjects. Namely, to specify a distance we can specify how long it would take a light beam to travel that distance. For Earth's radius, that computes to about one-fiftieth of a second. We say, therefore, that Earth's radius is equal to one-fiftieth of a light-second.

At higher rungs in the cosmic distance ladder it becomes

more practical to measure distances in light-years, rather than light-seconds. To get started with that, and for comparison purposes, let me record now that Earth's radius is roughly one-billionth of one light-year. Keep that tiny number in mind as we expand our survey of the world. It will soon encompass whole light-years, and then hundreds, millions, and finally billions of them.

Our next milestone length is the distance from Earth to our Sun. That distance is about 150 million kilometers, or 94 million miles. It is also 8 light-minutes, or about 15 millionths of one light-year.

Notably, the distance from Earth to the Sun is about 24,000 times Earth's radius. That startlingly large number emphasizes that even within the solar system, all of Earth, let alone a single human, really is "swallowed like a speck."

If you let such things bother you, be warned that it gets much worse. Our climb up the cosmic distance ladder has barely begun.

Knowing the size of Earth's orbit around the Sun, we can use it to determine the distance to some relatively nearby stars directly, using Euclidean geometry. Those stars are close enough that their position in the sky changes perceptibly over the course of the year, due to Earth's motion around the Sun. This effect is known as parallax. Our binocular vision uses parallax to gauge our distance to much nearer objects, which present different angles to our two eyes. The Hipparcos space mission, which operated from 1989 to 1993, used parallax to catalogue distances to about a hundred thousand (relatively) nearby stars.

The nearest star, Proxima Centauri, is a little over four light-years away. It has two nearby partners. Barnard's Star, the next nearest independent star, is about six light-years away. Communications with (hypothetical) extraterrestrials based in either of those systems, or with their future cyborg settlers, will require an abundance of patience.

Relative to interstellar space, our solar system is a cozy little den. The distance from our Sun to Proxima Centauri is about half a million times the distance from Earth to the Sun.

The key technique for extending the cosmic distance ladder still further exploits the fact, mentioned earlier, that we find the same sorts of objects and materials wherever we look. If we can identify a class of objects that all have the same intrinsic brightness, we say that those objects supply a "standard candle." If we know the distance to one realization of a standard candle, we can determine the distance to any other, simply by comparing the brightness we observe. For example, if one such source is twice as far away as another, then it will *appear* one-fourth as bright.

Now, all this begs the question of how we can convince ourselves that objects seen at different faraway places would have the same brightness if we got up close. The basic idea is that we look for classes of objects that have many properties in common, hope for the best, and check for consistency. A simple example will illustrate the basic idea, and its pitfalls.

Stars in general are much too diverse to serve as standard candles. White-hot Sirius A is about twenty-five times brighter than our Sun, while its nearby companion Sirius B, a dwarf star, is about one-fortieth as bright, even though both are—

astronomically speaking—roughly equally distant from Earth. We can do much better by restricting our comparisons to stars that have the same color—or, more precisely, stars that emit the same electromagnetic spectrum.* When we compare such otherwise identical-looking stars it is reasonable to hope that the difference in their brightness arises from a difference between their distances. The physical theory of stars, which explains many of their observed features, predicts this. But how can we check? One way is to find a compact group containing many stars close to one another. The Hyades cluster, which contains many hundreds of stars, is a prime example. If stars with very similar spectra have very similar intrinsic brightness, then two such stars that are within the same cluster should appear equally bright. And that's basically what we find.

Professional astronomers need to take several other complications, such as the effect of interstellar dust, into account. This dust, by absorbing light, can make objects appear more distant than they are. I hope my colleagues will excuse me for gliding over that and many other technicalities, which don't change the central idea.

We can extend our cosmic distance ladder, and "climb" from nearby objects to the limits of the visible universe, by using a variety of standard candles. Some kinds work better for relatively nearby objects, others for very distant ones. We must also check that they yield mutually consistent results.

The Hipparcos catalogue, mentioned earlier, gives us solid footing for our next step up the cosmic distance ladder. Having

* We can put this more poetically: They are stars that cast the same rainbows, up to overall brightness.

learned that similar stars have the same intrinsic brightness, we can use them to get the distance to more distant clusters, which are too far away to show observable parallax.

In this way, we can survey our own galaxy, the Milky Way. We discover that the stars in the Milky Way define a fairly flat disc, with a bulge in the middle. And we measure that the Milky Way is about a hundred thousand light-years across.

Cepheid variables are bright stars that pulsate. By careful study of Cepheid variables in the Magellanic Clouds,* Henrietta Leavitt (1868–1921) established that Cepheid variables that pulsate at the same rate also have the same brightness, and so provide standard candles. Cepheid variables are relatively easy to spot, because they are unusually bright as well as uniquely variable. Using Cepheid variables as their standard candles, astronomers have measured our distance to many galaxies.

Galaxies are distributed irregularly, so there is no unique value of the distance between them. Still, we can identify a typical distance between a galaxy and its nearest large neighbor. That intergalactic distance turns out to be a few hundred thousand light-years. Unlike the situation for stars or planets, which almost invariably are separated from their neighbors by distances many times their own size, the typical separation between galaxies is not vastly larger than the galaxies themselves.

There are several other useful standard candles, and many more interesting details of structure, within the realm of galaxies. Those riches of astronomy add depth to the picture I've

* The two Magellanic Clouds are minor galaxies that neighbor our own Milky Way. Prominent features of the Southern Hemisphere sky, they were used by Polynesian navigators long before Magellan.

sketched so far and reinforce its basic messages. But since my goal is to convey fundamentals, rather than to provide encyclopedic coverage, let us proceed to the farthest frontiers without further ado.

The Cosmic Horizon

In his pioneering studies of distant galaxies, using Cepheid variables as his primary tool, Edwin Hubble (1889–1953) discovered something fundamentally new, and rich in consequences. He observed that the patterns of starlight that distant galaxies emit—their spectra—are shifted toward longer wavelengths, compared with the light patterns of closer galaxies. This is called a redshift. The reason for that name is that if you systematically expand the wavelengths in a rainbow's light, then the colors of its stripes will change. Colors on the blue side will shift toward colors on the red side. This effect continues beyond what is humanly visible: A "new" blue stripe will appear where ultraviolet was before, and the red stripe will fade out into infrared.

Hubble's redshift observations have a compelling interpretation, which revolutionized our picture of the universe. The interpretation relies on a simple but striking effect, first described by Christian Doppler in 1842. Doppler pointed out that if a source of waves is moving away from us, then successive peaks in the wave pattern it emits will come from farther away, so that the waves will arrive stretched out. In other words, the observed waves will be shifted toward longer wavelengths than they would have had were the source stationary. The straightforward interpretation of Hubble's redshifts, therefore, is that they indicate the galaxies are moving away from us.

Hubble discovered a strikingly simple pattern within the redshifts he observed: The farther the galaxy, the larger the redshift. In more detail, he observed that the size of the redshift is proportional to the distance. This means that *the distant galaxies are moving away with speeds proportional to their distance.*

If we imagine reversing the galaxies' motions to reconstruct the past, then that proportionality acquires dramatic new meaning. It means that in the reversed flow, the more distant galaxies will be moving toward us more rapidly, covering the distance in just such a way that *everything comes together at the same time.* Thus, we are led to suspect that in the past all the matter in the universe was packed together much more tightly than it is today. Switching back to the original direction of time, it looks like a cosmic explosion.

Might the universe have emerged with a bang? When the Jesuit priest Georges Lemaître first proposed that interpretation of Hubble's observations, his "big bang" was a bold and beautiful idea, but the evidence for it was skimpy, and it lacked a firm basis in physics.* (Lemaître himself spoke of "the primeval atom" or "the cosmic egg." The less poetic name "big bang" came later.) But subsequent research has given us a much better understanding of matter in extreme conditions. Today, the evidence for the big bang concept is overwhelming. In chapter 6, we'll discuss cosmic history much more deeply, and review that evidence.

---

* Lemaître's basic theoretical work predated Hubble's observations.

Here, to round off our survey of the cosmos, we'll use the big bang picture to define the limit and extent of the visible universe. Running the movie of cosmic history backward in our minds, we found the galaxies all coming together to meet at a definite time. When did it happen? To calculate how long ago, we simply divide the distance a galaxy must travel by the speed at which it's moving. (Since a galaxy's speed is proportional to its distance, according to Hubble's observations, we'll find the same result consistently whichever galaxy we choose.) Doing that, we estimate that galaxies were all smushed together about 20 billion years ago. More accurate calculations, which include how the velocities change over time due to gravity, yield a somewhat smaller result. Today's best estimate is that 13.8 billion years have elapsed since the big bang.

When we look out to objects in the distant cosmos, we are looking at their past. Because light travels at a finite speed, the light from a distant object that we receive today had to be emitted long ago. When we look back 13.8 billion years or so, all the way to the big bang, we reach the limits of our vision. We get "blinded by the light." The initial cosmic explosion was so bright that we can't see beyond it. (At least, no one knows how.)

And because we can't see beyond a certain time, so, too, we can't see beyond a certain distance—namely, the distance that light can travel in the limited time available. However large the universe "really" is, the presently *visible* universe is finite.

How large is it? Here is where the idea of measuring distance in terms of light-years really shines. Since the limiting time is 13.8 billion years, the limiting distance is . . . 13.8 billion light-years! To bring the immensity home, let us recall here that Earth's radius is about one-billionth of one light-year.

With that wild contrast, our survey of cosmic size is complete. The world is large. There's plenty of room for humans to thrive in, and plenty left over for us to admire from a distance.

INNER PLENTY: WHAT WE KNOW AND HOW WE KNOW IT

Now let us look within. There, too, we will discover abundance. We will find again that there's plenty of space we can use, and much more that we can admire.

Different kinds of microscopes open our eyes to the riches contained in small things. Microscopy is a vast subject, full of ingenious and useful ideas. But here I will only briefly sketch four basic techniques, which reveal different levels of the deep structure of matter.

The simplest and most familiar microscopes exploit the ability of glass and some other transparent substances to bend light. By sculpting glass lenses and deploying them strategically, one can bend incoming light rays so as to spread out the angles with which they arrive at an observer's retina or a camera's plate. That makes the incoming image appear larger. This trick provides a wonderfully powerful and flexible way to explore the world down to distances of about a millionth of a meter, or a little less. (A meter is about 40 inches.) Using it, we can see the cells from which plants, animals, and humans are made. And we can glimpse the bacterial hordes, which both help and plague them.

When we push this light-bending technique to try to resolve even smaller objects, we run up against a fundamental problem. The technique is based on manipulating the paths of light rays. But the idea that light is composed of rays is only approximately right, since light travels in waves. Using waves to pick out details that are smaller than the size of the waves themselves is like trying to pick up a marble while wearing boxing gloves. The wavelengths involved in visible light are approximately half of one-millionth of a meter, so microscopes based on imaging visible light get fuzzy below that distance.

X-rays have wavelengths a hundred or a thousand times smaller than visible light, so they allow, in principle, access to much shorter distances. But there is no equivalent for x-rays of what glass supplies for visible light, namely a material that we can sculpt to make lenses and manipulate rays. Without lenses, the classical methods for magnifying images can't get off the ground.

Fortunately, there's a radically different approach that works. It is called x-ray diffraction. In x-ray diffraction, we dispense with lenses. We shine an x-ray beam on the object of interest, let the object bend and scatter the beam, and record what comes out. (To avoid confusion, let me note that these are quite different from the more familiar, simpler sort of x-ray images that doctors and dentists use. Those are much cruder projections, basically x-ray shadows. X-ray diffraction uses much more carefully controlled beams and smaller target samples.) The "pictures" that an x-ray diffraction camera takes don't look like that object at all, but they contain a lot of information about its shape, in a coded form.

A long and fascinating saga, strewn with Nobel Prizes, hinges on that qualification "a lot of." Unfortunately, x-ray diffraction patterns don't supply enough information to let you reconstruct the object purely by mathematical calculation. They're like corrupted files of digital images.

To address that problem, several generations of scientists constructed an *interpretive ladder*, which allows us to climb from simple objects to more complicated ones. The first objects to be deciphered from their x-ray diffraction patterns were simple crystals, starting with table salt. In that example, people had a good idea, based on chemistry, of what the answer should look like, namely a regular array of equal numbers of two kinds of atoms, sodium and chlorine. They also had reason to expect, based on the observed form of large salt crystals, that the array should be cubical. They did not know, however, the distance separating the atoms. Fortunately, you can calculate what the x-ray diffraction pattern looks like for model crystals with any possible value for the distance. By finding a match to the observed pattern, you both validate the model and determine the interatomic distance.

As scientists geared up to study more complicated materials, they used a kind of bootstrap procedure. At each stage, they used previously validated models to build up more elaborate models as candidates to describe materials with more elaborate spatial structures. Then they compared x-ray diffraction patterns calculated using the candidates to the ones they observed. Through a combination of inspired guesswork and heroic labors, success was sometimes achieved. With each new success structural features emerged, which could be fed as input into the next generation of models.

Historical highlights from this line of work include the extraordinary chemist Dorothy Crowfoot Hodgkin's determination of the three-dimensional structures of cholesterol (1937), penicillin (1946), vitamin B_{12} (1956), and insulin (1969), and the determination of the three-dimensional structure of DNA (1953)—the famous double helix—by Francis Crick and James Watson, who decoded x-ray diffraction pictures taken by Maurice Wilkins and Rosalind Franklin.

Today's much more advanced computers, using programs that incorporate the successful work of the past, allow chemists and biologists to solve more complicated x-ray diffraction problems routinely. In this way, they've determined the structure of tens of thousands of proteins and other important biomolecules. The art of scientific image manufacturing remains a vital frontier of biology and medicine.

To me, the interpretive ladder is both a beautiful example of and a metaphor for how we construct our models of the world more generally. In natural vision, we must turn the two-dimensional patterns that arrive at our retinas into a useful rendition of the three-dimensional world of objects in space. Abstractly, it is an impossible problem—there simply is not enough information. To compensate, we add assumptions about how the world works. We exploit abrupt changes in patterns of color, shadows, and motion to identify objects, their properties, their motion, and their distances.

Babies, or blind people suddenly given vision, have to learn how to see. They learn by experience to work with what they've got, building up from simple cases to construct a world that makes sense. Learning to "see" an object in its x-ray diffraction pattern has been a collective effort to accomplish something very similar; that is, to find a bag of tricks that lets us make sense of the world.

Our third technique, scanning microscopy, is refreshingly direct. One holds a needle with a tiny tip close to a surface of interest and "scans" by moving the tip parallel to the surface. If one does this while applying an electric field, then electric currents flow from the surface into the needle. The nearer the tip is to the surface, the larger the current. In this way, one can read out the topography of the surface with subatomic resolution. In images that reflect this data, one sees individual atoms towering up like mountains above a flat landscape.

Finally, let's discuss how scientists probe the smallest distances. The first experiment to get a look *inside* atoms was done by Hans Geiger and Ernest Marsden in 1913, with Ernest Rutherford guiding the effort. In their experiment, Geiger and Marsden pointed a beam of alpha particles at a gold foil. Some of the alpha particles were deflected by the foil. Geiger and Marsden counted how many got deflected through different angles. Before doing the experiment, they expected that few, if any, particles would be deflected by large angles. The alpha particles have a lot of inertia, so only close encounters with much heavier objects can change their course significantly. If the mass of the gold foil were spread out evenly, then large deflections wouldn't happen.

What they observed was quite different from their expectations. There were, in fact, significant numbers of large-angle deflections. Occasionally, alpha particles even reversed direction, returning back the way they came. Rutherford later recalled his reaction to the news:

> It was quite the most incredible event that has ever happened to me in my life. It was almost as incredible as if you fired a 15-inch shell at a piece of tissue paper and it came back and hit you. On consideration, I realized that this scattering backward must be the result of a single collision, and when I made calculations I saw that it was impossible to get anything of that order of magnitude unless you took a system in which the greater part of the mass of the atom was concentrated in a minute nucleus. It was then that I had the idea of an atom with a minute massive centre, carrying a charge.

Rutherford's detailed analysis of the Geiger-Marsden observations gave birth to the modern picture of atoms. He showed that to account for the data one must assume that most of the mass and all of the positive electric charge in an atom is concentrated in a tiny nucleus. Further refinements made those conclusions quantitative. An atomic nucleus contains more than 99 percent of the atom's mass. Yet a nucleus extends less than one-hundred-thousandth of its atom's radius and—being nearly spherical—occupies less than one part in a million of one part in a billion of its volume. Those are literally astronomical numbers. The way a nucleus is dwarfed by its atom parallels how the Sun is dwarfed by its surrounding interstellar space.

The Geiger-Marsden experiment established a paradigm for exploring the subatomic world that has dominated experimental work on fundamental interactions ever since. By bombarding targets with particles of ever-higher energy, and studying the patterns of their deflections, we learn about the targets' interiors. Here, too, we construct an interpretive ladder, using our understanding of what's revealed at each stage to design and interpret new experiments that probe deeper.

THE FUTURE OF SPACE

Beyond the Horizon

We can't see beyond the distance that light has traveled since the time of the big bang. That defines our cosmic horizon. But with each passing day the big bang recedes farther into the past. Space that was beyond our horizon yesterday comes within it today, and is newly opened to view.

Of course, since adding one more day, or even thousands of years, increases the age of the universe by only a small fraction, the fractional growth in the visible universe is hardly noticeable on human time scales. But it is entertaining to consider what kind of universe our distant descendants might perceive and to exercise our minds by thinking about what might be happening beyond the horizon. As Tennyson has Ulysses say:

> . . . all experience is an arch wherethrough
> Gleams that untraveled world whose margin fades

For ever and forever when I move.
How dull it is to pause, to make an end. . . .

The expanding cosmic horizon poses many questions. For instance, as the horizon expands, might the entire universe come within it? If space is finite, that will eventually happen. Famously, finite space need not have an edge. A sphere—that is, the surface of a ball—is an example of a space that is finite yet has no boundary. The surfaces of ordinary balls are two-dimensional. Though they are challenging to visualize, for mathematicians it is child's play to define three-dimensional spaces that, like ordinary spheres, are finite yet have no boundary. Such spaces provide candidate shapes for a finite universe.

The visible universe is remarkably uniform. It contains the same kinds of materials, obeying the same laws, organized in the same ways, evenly distributed throughout. Another question raised by the expanding horizon is whether or not that "universal" pattern holds up for the parts we can't see yet.

Or is the universe really a "multiverse," home to many different patterns or laws? The most straightforward way to answer this question would be to observe outlandish things happening far away. Were that to happen, we could establish the multiverse experimentally. A sad but perfectly logical possibility is that other facts about the fundamental laws and cosmology will suggest that we live in a multiverse, but will also suggest that the "different" parts will become visible only in the very distant future, when the horizon expands to contain them. I call this possibility sad, because to me using an idea to say something concrete about the world we experience brings it to another level. It's where the magic is. Also, testing keeps you honest.

Particles of Space?

Euclid assumed that one could continue to measure distance more and more finely, without limit, using the same conceptual tools. He didn't know about atoms, elementary particles, or quantum mechanics. Now we know better. When we divide matter into very small parts, things change a lot! A placid drop of water, which appears continuous and at rest, breaks up into atoms and even more basic units, which jitter and jive to the tune of quantum mechanics.

When we come to measure subatomic distances, we must use tools that are very different from the sorts of rigid rulers Euclid had in mind. Scalable versions of those instruments simply don't exist. Yet Euclid's geometry lives on, triumphant, within our fundamental equations. Within those equations, elementary particles (and the fields that support them) occupy a seamless continuum, equivalent in all its parts, measured in lengths and angles, obedient to Pythagoras's theorem, just as Euclid assumed. It's uncanny that Nature has let us get away with it. So far . . .

. . . but probably not forever. According to Einstein's general theory of relativity, space is a kind of material. It is a dynamic entity, which can bend and move. In our later discussions, many other reasons to consider space as a material will emerge as well. According to the principles of quantum mechanics, anything that can move does move, spontaneously. As a result, the distance between two points fluctuates. Upon combining general relativity with quantum mechanics, we calculate that space is a kind of quivering Jell-O, in constant motion.

When the distance between two points is not too small, those quantum fluctuations in distance are predicted to be a negligible fraction of the distance itself. Then we can ignore them, as a practical matter, and return to the comfort of Euclidean geometry. But when we refine our focus below about 10^{-33} centimeters—a tiny distance known as the Planck length—then typical fluctuations in the distance between two points can be as large or larger than the distance itself. Two lines from William Butler Yeats's apocalyptic vision spring to mind:

> . . . the center cannot hold;
> Mere anarchy is loosed upon the world. . . .

Writhing rulers and dancing compasses undermine the foundations of Euclid's approach to geometry, and ultimately Einstein's, too. The ideas of GPS can't be scaled down, because the orbits of satellites in Planck-length detail are noisy and unpredictable. What will replace them? Nobody knows for sure. There's little prospect of guidance from experiment, because the Planck length is thousands of trillions times smaller than distances we know how to resolve. For me, though, it is difficult to resist the idea that space-time is not essentially different from matter, which we understand much more deeply. If so, it will consist of vast numbers of identical units—"particles of space"—each in contact with a few neighbors, exchanging messages, joining and breaking apart, giving birth and passing away.

2

THERE'S PLENTY OF TIME

PRELUDE: MEASURE AND MEANING

Frank Ramsey (1903–1930) blazed bright, though briefly. Before dying of liver problems at the age of twenty-six, Ramsey made seminal contributions to mathematics, economics, and philosophy. Despite his youth he was a central figure in intellectual life at Cambridge in the 1920s. He collaborated and argued with both John Maynard Keynes and Ludwig Wittgenstein, who are widely regarded as the greatest economist and the greatest philosopher of the twentieth century, respectively, on their home turfs. "Ramsey theory" is a thriving, entertaining corner of mathematics that grew out of his work.

(Here's a classic little gem that will give you a taste of Ramsey theory: Among any group of six people, each pair of whom are either friends or enemies, there will either be a set of three people who are all mutual friends, or a set of three people who are all mutual enemies.)

Frank Ramsey is a thinker to be reckoned with. His objection to the significance of the physical world's superhuman proportions deserves serious attention:

> My picture of the world is drawn in perspective and not like a model to scale. The foreground is occupied by human beings and the stars are all as small as three-penny bits. I don't really believe in astronomy, except as a complicated description of part of the course of human and possibly animal sensation. I apply my perspective not merely to space but also to time. In time the world will cool and everything will die; but that is a long time off still and its present value at compound discount is almost nothing.

A famous *New Yorker* cover expresses a similar thought. It shows a "map of the world" where most of the drawing is devoted to Manhattan while the rest of our planet gets squeezed into a cramped, sketchy background.

Ramsey's attitude is a healthy corrective to cosmic sizeism. Equal volumes of space are equal in their potential for accommodating matter and motion, but that does not mean they are of equal importance. The undifferentiated, empty regions are less interesting. Similarly, equal intervals of time are equal in their ability to accommodate the ticking of clocks, but that does not mean they are of equal importance. To most of us, most of the time, nearby events matter more. It is an attitude that comes naturally to us, as children, as a strategy to cope in the world.

But Ramsey, in retaining that attitude, takes it too far. When he says he does not believe in astronomy, I do not believe him. His statement hints to me instead that the outrageous hugeness of cosmic space and time bothered him deeply, as it had bothered Pascal. Sadly, by denying their significance, he cut himself off from a potential source of inspiration. He missed the opportunity to become a great cosmologist, as well as mathematician, economist, and philosopher.

We can recognize both that there's plenty "out there" and that there's plenty "in here." Neither fact contradicts the other, and we do not have to choose between them. From different perspectives, we are both small and large. Both perspectives capture important truths about our place in the scheme of things. To get a full and realistic understanding of reality, we must embrace them both.

TIME'S ABUNDANCE

As we find for space, so it is also for time: There's plenty of it, both outside and inside. Though the immensities of cosmic time dwarf us, yet we contain immensities of time within.

In his visionary cosmic history *Starmaker*, Olaf Stapledon, a pioneering genius of science fiction, writes, "Thus the whole duration of humanity, with its many sequent species and its incessant downpour of generations, is but a flash in the lifetime of the cosmos." The Roman philosopher Seneca expressed the opposite of this thought in "On the Shortness of Life." "Why do we complain about nature?" he writes. "It has acted generously; life, if you know how to use it, is long."

As we'll see, both Stapledon and Seneca got it right.

WHAT IS TIME?

Lest we drown in vagueness and nonsense, let us pause to take a deep breath and address a very basic question: What is time?

Time seems, as a matter of psychology, less tangible than space. We can't move around freely in time, or even revisit a chosen moment. Once a moment is passed, it is past. It is not, then it is, then it is not again.

Saint Augustine, a very powerful thinker, articulated a common feeling of puzzlement: "What then is time? If no one asks me, I know what it is. If I wish to explain it to him who asks, I do not know."

One witty but unserious answer often has been misattributed to Einstein, though it originates from the science-fiction writer Ray Cummings: "Time is what prevents everything from happening at once."

Another pithy response, which at first may sound no more serious, is that "time is what clocks measure." Yet that, I believe, is the germ of the correct answer. It is the thought we'll build on.

There are many phenomena in nature that repeat regularly. The cycles of day and night, of the Moon's waxing and waning, of the seasons, and of the beating of human and animal hearts are obvious features of common experience. If we compare the rate of one person's resting heartbeat with another's, we find that roughly the same ratio persists over many beats. We find, too, that each cycle of lunar phases—each lunar month—contains very nearly the same number of days.

The cycle of seasons seems hazier, on the face of it, due to the vagaries of weather. To refine their predictions of seasons, people in several civilizations developed a technology of astronomical timekeeping. They hit on the idea of monitoring changes in the path of the Sun's march across the sky—where it rises, where it sets, and how high it rises, day by day. The changes in those positions are much more regular than seasonal changes in weather patterns, which fluctuate unpredictably. By monitoring the Sun, people achieved a much more precise and useful definition of seasons and years. (Seasons are officially defined as intervals between solstices, which mark the most extreme solar excursions, and equinoxes, which mark the most rapid daily changes. Solstices also mark the extreme divisions of day and night, while equinoxes mark their equality. Years are the intervals that pass between complete cycles of change.) Having made those precise definitions, people observed that each season contains the same number of days, or of lunar months, year after year. They constructed calendars, which helped them in many aspects of life, such as deciding when to plant crops, anticipating when they'd need to harvest, and, for hunters, when to expect animal migrations.

In short, we find that many different cyclical processes, physiological and astronomical, are synchronized. They march to the same drummer. We can use any of them to measure the others.* The observation that there is a shared, universal pace is a deep fact about the way the physical world works. To express the pace itself, we say that there is something that all the world's cycles tap into, which tells them when to repeat. That something, *by definition*, is time. Time is the drummer to which change marches.

Two other manifestations of time are central to human experience. One is its role in music. In playing music together, or in dancing or singing, we rely on our expectation that everyone involved will stay in sync. While that experience is so familiar that we tend to take it for granted, it provides convincing evidence that we share, with high accuracy, a common notion of the passage of time.

Another manifestation of time, perhaps the most important of all for humans, relates to life history. Almost all babies develop on roughly the same schedule, beginning to walk, talk, and achieve other milestones after a certain number of months (or days or weeks). People grow in height, reach puberty, thrive, and decline according to predictable patterns, closely connected to the number of years they've lived. Each of us is a clock, albeit one that's hard to read accurately.

As the arc of human life history illustrates, time controls the progress of noncyclical events, as well as cyclical ones. As people became scientifically sophisticated and studied motion and other kinds of change in the physical world systematically, they found again and again—in every case, so far—that

---

* To be sure, it would take great patience to measure days in heartbeats. But one can use the progress of shadows, for example, to divide the day more finely.

all change proceeds according to a common rhythm. Changes in the positions of astronomical bodies, changes in the positions of bodies in response to forces, the unfolding of chemical reactions, the progress of light beams through space—those changes, and many more, all evolve to the tempo of a single time.

Putting it another way: There is a quantity, usually written as *t*, which appears in our fundamental description of how change takes place in the physical world. It is also what people are talking about when they ask, "What time is it?" *That* is what time is. Time is what clocks measure, and everything that changes is a clock.

HISTORICAL TIME: WHAT WE KNOW AND HOW WE KNOW IT

We took the measure of cosmic time already, in the preceding chapter, when we looked back to the big bang. Since then, 13.8 billion years have passed. On the scale of human longevity, that is a very long time, indeed. It encompasses hundreds of millions of human lifetimes.

It is a mind-boggling figure, 13.8 billion years, but the big bang is remote from our experience. To appreciate the abundance of time, we should also consider deep history closer to home. There are two approaches to measuring very long times: radioactive dating and stellar astrophysics. Let's discuss them in turn.

Radioactive dating is based on the existence of nuclear isotopes. These are atomic nuclei that contain the same number of protons but different numbers of neutrons. Such nuclei give rise to atoms that have nearly identical chemical properties. But many kinds of atomic nuclei are unstable, and decay, each with a characteristic lifetime. Often isotopes of the same chemical element have radically different lifetimes. Those two features—same chemistry, different lifetimes—are what we exploit to do radioactive dating.

To keep things concrete, let's focus on one important example of radioactive dating, which uses carbon. The most common isotope of carbon is ^{12}C ("carbon-12"), which contains six protons and six neutrons. ^{12}C nuclei are highly stable. But there is also another significant isotope of carbon, ^{14}C ("carbon-14"), which is unstable, or "radioactive."

^{14}C has a half-life of about 5,730 years, meaning that if you have a sample of material containing ^{14}C atoms, in 5,730 years half of them will be gone. What happens is that the ^{14}C nuclei convert into nitrogen nuclei (^{14}N) while emitting electrons and antineutrinos. We'll be discussing processes of this sort—radioactivity and the weak force—more deeply later. For present purposes, the details aren't crucial.

Of course, we don't have to wait 5,730 years to check that picture out. Because even small samples of organic matter contain many carbon atoms, we can detect many decays within small intervals of time. What we observe, when we monitor the outflow of electrons, is that in equal intervals of time an equal proportion of the surviving ^{14}C nuclei decay.

Since the universe is much older than 5,730 years, the question arises: Why is any ^{14}C left? The key fact here is that new

^{14}C nuclei are being created in Earth's atmosphere, through the action of cosmic rays. That creation compensates for the decays and maintains a balance between ^{14}C versus ^{12}C in the atmosphere.

Living things take in carbon either directly from the atmosphere or shortly after it dissolves from the atmosphere into water. The carbon they ingest reflects the current atmospheric $^{14}C/^{12}C$ balance. But once it is incorporated into their bodies, the decaying ^{14}C is no longer replenished. After that its fraction decreases with time, in a predictable way. Thus, by measuring the ratio of ^{14}C to ^{12}C in a sample of biological origin, one can determine when the source of the sample was last alive and capturing carbon.

There are two practical ways to measure the ratio. Since there are always far more ^{12}C nuclei than ^{14}C, we can get a good estimate of ^{12}C abundance simply by weighing the total carbon. To get the ^{14}C abundance, we can measure the radioactivity—that is, the rate of electron emission. Since we know what proportion of ^{14}C decays in an interval of time, we can leverage that measurement to infer the ^{14}C content.

A more modern method is to take the sample to an accelerator, where you can physically separate the ^{14}C and ^{12}C, by exploiting their different motions in strong electric and magnetic fields. The two methods yield consistent results.

Carbon dating is widely used in archaeology and paleobiology. It has been used to date ancient Egyptian and Neanderthal artifacts, for example, including mummies. We can check some of those Egyptian artifacts against historical records, and we find agreement. The Neanderthals didn't keep records, but thanks to carbon dating we know that they flourished in Europe for several hundred thousand years, and as recently as forty thousand years ago.

We can also date bones and artifacts of early modern humans (*Homo sapiens*). From those remains, we infer that our species has been around for about three hundred thousand years. The early record is sparse, indicating that populations were small: *Homo sapiens* was not a particularly successful species early on.

It is important to emphasize that there are many ways to validate the ages obtained by carbon dating. We can construct a time ladder, similar in spirit to the distance ladders we discussed earlier. A simple, classic, and particularly beautiful example involves old trees. Trees add a ring to their bark each year, as the wood deposited during different seasons looks different, providing contrast. We can check that carbon dating reproduces the correct relative ages for the different bands, as well as yielding the overall age.

There are many other isotope pairs besides carbon ^{14}C and ^{12}C, with a wide range of half-lives. Using essentially the same techniques, we can use them to measure much longer times than carbon dating reaches. For example, isotopes of uranium and lead have been used to obtain the age of mineral samples (gneiss) from western Greenland. They give concordant ages in the neighborhood of 3.6 billion years. Thus, we infer that those rocks formed 3.6 billion years ago, and have undergone little chemical processing since. In this way, we learn that Earth has existed as a solid planet for a significant fraction—more than a quarter—of the lifetime of the visible universe.

The astrophysical theory of stars suggests a method to determine their ages. Stars generate energy by burning nuclear fuel. As the fuel is consumed, they change their size, shape, and color. Our Sun, for example, is predicted to become a red giant in about five billion years. Then its body will consume Mercury and Venus, and things will get quite nasty on Earth. Roughly a billion years later, according to theory, the Sun will blow off its extended atmosphere and settle down into a hot, Earth-sized white dwarf. Then it will slowly cool and eventually, over several billion years, fade to black.

There are many ways to test the theory of stellar evolution. For example, we can look at groups of stars that gather closely together in a cluster. It is reasonable to think that many of those stars will have formed at roughly the same time (on cosmic scales). If so, then they should all have the same age. As stars age, they evolve in predictable ways, changing their color and brightness. Using the theory of stellar evolution, we can compute the age of each star separately. Astronomers have found in many cases that the computed ages within a cluster do agree with one another, thus both vindicating the theory and dating the cluster.

We find in this way that some of the oldest stars are almost as old as the visible universe. In other words, star formation commenced within one or two billion years after the big bang. On the other hand, some stars are quite young, and we also observe regions where stars are still forming.

Summarizing, we can say that:

- The universe commenced forming stars and planets quite early in its history, about thirteen billion years ago. New stars continue to form, though at a diminishing rate.
- The Sun and Earth have been around in something close to their present form for about five billion years.
- Humans have been around in something close to their present form for a much briefer time, about three hundred thousand years. This amounts to about ten thousand generations, or five thousand human lifetimes.

INNER TIME: WHAT WE KNOW AND HOW WE KNOW IT

The *inner abundance* of time appears when we compare the span of a human lifetime with the speed of the basic electrical and chemical processes that enable thought. That comparison reveals that a lifetime can support immensities of individual experiences and insights.

The Speed of Thought

Wolfgang Amadeus Mozart died at thirty-five years of age; Franz Schubert at thirty-one; Évariste Galois, the great mathematician, at twenty; James Clerk Maxwell, the great physicist, at forty-eight. Evidently, it is possible to squeeze a lot of creative thoughts into a human lifetime. How many?

No single measure of speed applies to the bewildering variety of brain processes, so there is some vagueness in the question. Still, I think it is possible to give a rough but meaningful answer.

One fundamental limitation to human signal processing is the downtime (latency) between the pulses of electrical activity (action potentials) that neurons use to communicate with one another. This recovery period limits the number of pulses to a few tens or hundreds per second, depending on the neuron type. It is probably no accident that the "frame rate" at which we can start to distinguish that movies are actually a sequence of stills is about forty per second, just adequate to accommodate a modest number of pulses. That frame rate is an objective measure of how fast we can process visual signals into forms that our brains can make use of. It means that we process, and "understand," about a hundred billion distinct scenes in a lifetime.

The number of conscious thoughts we can entertain is probably significantly less than that, yet still enormous. Average speech rates, for example, are about two words per second. If we estimate that five words represent a significant thought, then a lifetime has room for about a billion thoughts.

Those estimates testify that we're gifted with over a billion opportunities to experience the world. In that important sense, there's plenty of inner time. That estimate might even be too conservative, since our brains support parallel processing, whereby several different thoughts can be running—mostly subconsciously—at once.

T. S. Eliot, in "The Love Song of J. Alfred Prufrock," had

a more ironic take on the same conclusion: "In a minute there is time / For decisions and revisions which a minute will reverse."

Helped by our ancestors and our machines, we can augment our thought resources greatly. We need not rediscover from scratch how to fulfill basic needs like staying warm or obtaining food and drink. On a more elevated plane, we need not rediscover calculus, or the foundations of modern science and technology. Nor, thanks to modern computers and the internet, need we spend precious thought cycles on laborious calculations, or on memorizing masses of information. By bringing in those helpers, we can outsource immensities of thinking and free up more of our internal time for other uses.

Nature is not limited by the speed of human thought. Events can happen much faster than our processing rate of forty per second, even though our vision can't resolve them. Notably, the "clock rate" for modern information processors, such as the CPU of a high-end laptop, is approaching 10 gigahertz, corresponding to ten billion operations per second. Computers can work much faster than brains, because transistors use the electrically driven motion of electrons, instead of the much slower processes of diffusion and chemical change that neurons rely on. By this natural measure, the limiting speed of thought for artificial intelligence is roughly a billion times faster than the speed of thought for natural intelligence.

MEASURING TIME

The history of clocks and the measurement of time brings in much of the history of physics. Early clocks include instruments based on the position of the Sun (sundials), hourglasses based on the flow of sand, and related devices based on the flow of water, candles, and others. Legendary figures such as Galileo and Christian Huygens developed mechanical pendulum clocks, which were improved over many decades, and set the standard for accuracy until well into the twentieth century.

The twentieth century brought in more reliable clocks, based on entirely different physical principles. At the frontier of clock making, swinging pendulums and unwinding springs got replaced by vibrating crystals, and then by vibrating atoms. Those smaller oscillators are less exposed to buffeting from the external world, and they operate with very little friction. As a result, today's most accurate atomic clocks are extraordinarily stable—within a part in 10^{-18}, to be precise. Thus, two such clocks, operating over the span of the lifetime of the universe, would continue to agree within about one second. Today, relatively cheap, compact (chip-size) atomic clocks can keep time with 10^{-13} accuracy. They gain or lose a few seconds every million years.

Those extraordinary accuracies might seem extravagant, but actually they are extremely useful. For one thing, they translate, in the Global Positioning System, to precise distance measurements. (Such measurements make it possible, for example, to align large machines precisely.) Note that even tiny errors in time, when multiplied by the speed of light, can translate into noticeable errors in distance.

The design of ever more precise and accurate clocks is a challenging, wonderfully creative branch of modern physics. A recent example is close to my heart: It may be possible to orchestrate large numbers of atoms, cooperating within a new state of matter that I predicted and that was subsequently observed—a "time crystal"—to improve on the accuracy of single-atom atomic clocks.

Resolving Short Times

Just as we discussed earlier for space, when we come to extremely short periods of time, we must measure it in different, less direct ways. In the spatial case, we saw that x-ray diffraction and scattering in the style of Geiger and Marsden gave information that could be converted into maps (that is, images) of the atomic and subatomic world. Those techniques involve observing how targets—namely, the objects we want to image—change the motion of incident x-rays and of incident particles, respectively.

To resolve the structure of rapid events, we employ methods of a similar kind, but focus on changes in energy rather than changes in direction of motion. The world of rapid events is full of wonders and surprises. Let me spotlight a couple of highlights—briefly, as befits the subject matter.

Using high-powered lasers, it is possible to resolve the sequence of events that occur in many chemical and biochemical processes. *Femtochemistry* constructs those timelines, in steps as small as 10^{-15} seconds (one femtosecond). With understanding, increasingly, comes control. Lasik eye surgery exploits femtosecond laser pulses to remodel patients' corneas.

It is possible to resolve even shorter times by using high-energy accelerators. We'll explore examples of this more deeply later. The Higgs particle, whose discovery was a major triumph for twenty-first-century physics, is highly unstable. It lives for only about 10^{-22} second. Thus, in order to discern evidence for its existence, physicists had to reconstruct events on that time scale.

THE FUTURE OF TIME

Engineering Physical Time

Einstein's theory of general relativity has gone from triumph to triumph as our theory of gravity. It teaches us that space-time can bend and distort. That fact helps to fuel dreams of time travel, portals, wormholes, and warp drives. Might those fantasies and desires become engineering realities?

I see little hope that we'll be able to manipulate physical time in the foreseeable future. Ironically, the Laser Interferometer Gravitational-Wave Observatory's (LIGO) observation of gravitational waves, which is the most recent major confirmation of general relativity, and perhaps the purest, also demonstrates the problem starkly.

LIGO is an exquisite instrument, designed to detect tiny distortions in space-time. It is sensitive to changes in the relative positions of mirrors, separated by four kilometers, that are a thousand times smaller than the size of an atomic nucleus. Yet even with that kind of sensitivity, LIGO was barely able to detect distortions produced by the violent merger of two black holes, each several times the mass of the Sun. The message is simple: Space-time *can* be distorted, *but* it's very hard work.

Engineering Psychological Time: Hopping and Cycling

Physical time is very stiff. For practical purposes it flows steadily and in one direction, the same for every entity in the physical universe. Psychological time is quite different. It can meander, branch, and jump around quite nimbly. We can revisit the past, consulting memory. In doing this, we can move through it quickly, or slowly, or in jumps. Or we can change it, by imagining how things might have been. We routinely imagine alternative futures and plan actions to realize desirable ones. That may be the central task of our frontal lobes—those massive, convoluted outcroppings of brain that uniquely distinguish humans among animals.

Computers are essentially ageless, and they can revisit previous states precisely, and they can pursue several programs in parallel. An artificial intelligence rooted in those platforms will be able to engineer its psychological time with great precision and flexibility. Notably, it could set up states that lead to pleasure, and relive them repeatedly, while experiencing each as fresh.

Engineering Psychological Time: Speed

There's a big gap between the human speed of thought—which we estimated at a few tens per second—and the existing speed of electron-motion-powered thought, as embodied in computer clock rates. It's about a factor of a billion, as we've discussed. Basic femtosecond atomic processes are even faster, by an additional factor of many thousands. Thus, there's room to pack a lot more life into each moment.

Artfully evolved humans, cyborgs, or completely artificial intelligences have plenty of room to transcend the (presently) standard speed of thought. Barring catastrophic nuclear war or climate change, that will soon come to pass—I'd guess within a few decades.

More fancifully, we can imagine forms of intelligence based on subatomic processes, which can be even quicker. Robert Forward's delightful hard sci-fi novel *Dragon's Egg* plays on that theme. He imagines an intelligent form of life, the cheela, evolving on the surface of a neutron star. There, nuclear chemistry rather than atomic chemistry would rule. Nuclear chemistry involves much larger exchanges of energy than atomic chemistry, and therefore operates faster. Epochs of cheela history pass in the blink of a human eye. The human astronauts who come upon a savage, scientifically backward form of life discover, half an hour later, that the cheela, given access to their libraries, have far outstripped them.

Engineering Psychological Time: Persistence

In *Gulliver's Travels*, Jonathan Swift introduced a race of immortals, the Struldbruggs. Though immortal, the Struldbruggs grow old. They become frail, miserable creatures that are a burden to society. The misery or evil of immortality is a common theme in myth and literature. The intended lesson: When it comes to longevity, be careful what you wish for.

Frankly, I think this is sour grapes. The destruction of memory and learning by death is horrifying and wasteful. Extending the healthful human lifespan should be one of the main goals of science.

3

THERE ARE VERY FEW INGREDIENTS

As children, we learn to deal with many sorts of things: other people, animals, plants, water, soil, stones, wind, the Sun and the Moon, stars, clouds, books, smartphones, and many others. We develop different models for how to identify each of those things, how they might affect us, and how we can affect them. The idea that all of those things are made from a handful of basic building blocks, each occurring in great numbers, is not an important part of those models. It is, however, a central lesson of science.

ATOMS AND BEYOND

> If, in some cataclysm, all of scientific knowledge were to be destroyed, and only one sentence passed on to the next generations of creatures, what statement would contain the most information in the fewest words? I believe it is the *atomic hypothesis* (or the atomic

> *fact*, or whatever you wish to call it) that *all things are made of atoms.*
>
> —*Richard Feynman*

The word "atom" derives from a Greek root meaning "without parts." For a long time, scientists thought that the smallest objects that can be exchanged in chemical reactions were the ultimate, indivisible units of matter. Those basic chemical building blocks got to be called "atoms," and that name has stuck.

But when people studied matter in more extreme conditions than are commonly encountered in chemistry, they discovered that chemical "atoms" can be broken into smaller units. Thus, the "atoms" of chemistry, which are the objects that go under that name in most of the scientific literature, are not "atoms" in the sense of being our ultimate building blocks.

The traditional atom of chemistry consists of electrons surrounding an atomic nucleus. The nucleus can be further analyzed into protons and neutrons. That's not the end of the story. Our best world-models today build up atoms from electrons, photons, quarks, and gluons. As we'll see, there are good reasons to think that this really *is* the last word.

These discoveries, which form part of our fundamentals, continue the spirit of the atomic hypothesis. They suggest that we should rephrase (and maybe rename) it, though. Instead of "all things are made of atoms," we should say that "all things are made of elementary particles." But whichever way you state it, the central message is clear: It pays to analyze matter into the smallest units you can. After doing that correctly, you can build back up, conceptually, and construct the physical world.

The modern scientific construction of physical reality from a few simple ingredients requires that we reimagine both what we mean by "simple ingredients" and how we do "construction." Our everyday experiences do not prepare us well for the modern versions of those concepts.

PRINCIPLES: REALITY AND ITS RIVALS

The most basic ingredients of physical reality are a few principles and properties. Those principles and properties are *expressed* through things we call elementary particles. But the "elementary particles" differ in important ways from any objects of common experience, and to understand them properly we must start with the principles and properties.

Four (Deceptively) Easy Principles

Four simple yet profound general principles govern how the world works. I'll first state them all at once, telegraphically, and then spell them out in more depth.

1. *The basic laws describe change.* It is useful to separate the description of the world into two parts: states and laws. States describe "what there is," while laws describe "how things change."
2. *The basic laws are universal.* That is, the basic laws hold everywhere, and for all times.
3. *The basic laws are local.* That is, the behavior of an ob-

ject in the immediate future depends only on current conditions in its immediate vicinity. The standard scientific jargon for this principle is *locality*.

4. *The basic laws are precise.* The laws are precise, and they admit no exceptions. Thus, they can be formulated as mathematical equations.

The simplicity of those general principles is deceptive. They are far from self-evident. They may not even be completely true. Their strength derives not from any logical necessity, but from their proven success. They have pointed us to an impressively successful description of how the physical world actually works, as this book aims to document.

Over the bulk of human history, people have held many different views about how the physical world works. Ideas that contradict one or more of our principles have been recorded in folklore, in history, and—until recently—in the works of learned academics, philosophers, and theologians. Some, such as astrology, telepathy, clairvoyance, and witchcraft, bring in forces that act powerfully across big separations in space and time. Others, such as extrasensory perception, telekinesis, prayer-induced miracles, and magical thinking, assign prominent roles in shaping the course of physical events to mind and will. Most of those ideas are "reasonable" extensions of the world-models we build up as children, in which our mind is disembodied and our will controls our body. Historically, most people's world-models have accepted many or all of them.

Only a tiny percentage of people over the course of human history have aspired to make precise predictions about what happens next under carefully controlled conditions, or even imagined that such a thing might be possible. Yet that possibility is the central message of our principles. Our general principles were first clearly formulated in the seventeenth century. They are the core lessons of the Scientific Revolution.

The message of the first principle is simply that "What happens next?" is a more approachable question, and proves to be a much more fruitful question, than "Why are things the way they are?" "What happens next?" is an approachable question because, thanks to our second and third principles, we can do experiments to answer it. That is, we can make an accurate copy of the situation we're interested in—set up the same state—and observe what happens in the copy.

A crucial point of the second principle, which helps make that "obvious" suggestion—to perform experiments—a practical one, is that we can do the experiments anywhere and at any time. According to the second principle, universality, we will always find the same fundamental laws.

The third principle, locality, allows another crucial simplification. It tells us that in formulating the laws, we do not need to take into account the whole universe, or all of history. It tells us, more precisely, that we can aspire to control all the relevant conditions by taking appropriate precautions here and now.

Finally, the fourth principle, precision, is an invitation to ambition. It says that if we describe the laws using appropriate concepts, we can get a description that is brief yet complete and fully accurate. It is also a challenge: We should not be satisfied with less.

In short, these principles assure us that by doing experiments, we can discover precise, universal laws that govern how things change. Science pursues that goal systematically, and relentlessly.

Principles one through four, working together, give us a strategy to make discoveries. We start by studying what happens in precisely defined, simple situations that we can set up repeatedly. Having mastered those, we can try to deduce what will happen in more complicated situations.

Babies—even animal babies—use that same experimental strategy to get in tune with physical reality. We humans learn, for example, how to throw a ball, how to bring food to our mouths, and hundreds of other practical procedures to make changes in the physical world by weaving together experiences at different times and places, under different conditions. Scientists, and people who open themselves to science, are born-again explorers. But we "babies" get to benefit in our explorations from logical minds, sense-enhancing instruments, and the work of explorers who came before us.

Newton and Locality

Newton was extremely unhappy with one of his most glorious discoveries. According to Newton's law, the gravitational force that one body (call it body B) exerts on another one (call it body A) acts immediately, with no delay in time, however far the two bodies are separated. This implies that you cannot predict the motion of A based solely on conditions in the immediate neighborhood of A—specifically, you have to know where B is. Newton was deeply dissatisfied with that feature of his own law, as he expressed in a letter to his friend Richard Bentley:

> That one body may act upon another at a distance through a vacuum, without the mediation of anything else, by and through which their action and force may be conveyed from one to another, is to me so great an absurdity that I believe no man who has in philosophical matters a competent faculty of thinking can ever fall into it.

Newton realized that his law of gravity is not local—in other words, that it fails to embody our third principle—and he did not like it.

This perceived flaw was, for Newton and for several generations of scientists who followed him, purely theoretical. Newton's law of gravity worked spectacularly well in practice. You might say its shortcomings were aesthetic, or even, for Newton himself, theological. It seemed to represent a lapse in God's usually excellent taste.

Newton's faith in our third principle—the principle of local action—proved amazingly prescient. Many decades after his death, starting in the mid-nineteenth century, physicists filled the passive "vacuum"—a nothingness, or Void—that Newton complained about with force-transmitting materials, which we call fields. Fields, rather than particles, are the fundamental building blocks of matter in modern physics.*

* Here, with Newton, we have anticipated a major theme of chapter 4.

A Case Study: Atomic Clocks

Atomic clocks are a superb example of our fundamental principles at work.

Vibrating atoms supply the heartbeat of atomic clocks. Their physical state determines how they change—in this case, how fast they vibrate (fulfilling the first principle). Importantly, experimenters have measured rates of atomic vibrations at different times and places, and always found consistent answers (fulfilling the second principle)—once they take a few laboratory precautions (exploiting, and fulfilling, the third principle). And, as we discussed previously, atomic rates of vibration have been measured with exquisite precision, with consistent results (fulfilling the fourth principle).

The trickiest part, both in this case and in most experiments, is taking "necessary precautions." To get consistent results, experimenters need to make sure that the complicated, finely tuned apparatus they use to trap atoms and observe their behavior—including lasers, fancy cooling equipment, vacuum chambers, and a lot of complicated electronics—is stable. You must shield it against the effects of ground-shifting tremors set in motion by passing trucks, and the seismic rumblings of Earth itself. You mustn't let playful children or heedless students wander through the lab, touching things. The point of the third principle—locality—is that these precautions, and other humdrum corrections for temperature, humidity, and so forth, all relate to local conditions. (The truck might be far away, but what counts are tremors at the lab itself.) Thankfully, you don't have to worry about the distant universe, what happened in the past, or what will happen in the future.

The heart of the matter is the atoms. What vagaries do we need to control for before we get to the reproducible, exquisitely precise results that atomic clocks are famous for? Basically, just four things. We need to keep the atoms of interest separated from other atoms. That's what the cooling apparatus and vacuum chambers are for. And we have to keep track of electric, magnetic, and gravitational conditions where the atom is—the value of the electric, magnetic, and gravitational fields, as we say. Those fields can be measured locally, by monitoring how charged particles move and how fast bodies fall. Once you've made appropriate corrections for that small list of local conditions, you've done enough. At that point, you will always observe a consistent rate of atomic vibration, with extremely high precision—or else you will have made a great discovery, which has eluded all previous experimenters!

It is philosophically important to notice that it is unnecessary to take into account what people, or hypothetical superhuman beings, are thinking. Our experience with delicate, ultra-precise experiments puts severe pressure on the idea that minds can act directly on matter, through will. There's an excellent opportunity here for magicians to cast spells, or for someone with extrasensory powers to show their stuff, or for an ambitious experimenter to earn everlasting glory by demonstrating the power of prayer or wishful thinking. Even very small effects could be detected. But nobody has ever done this successfully.

WHAT MIGHT HAVE GONE WRONG, BUT HASN'T

Before concluding our discussion of world-building principles, I will show, using a simple thought experiment, how our principles could have been wrong. In fact, I'll describe plausible universes of the future in which our principles *won't* hold.

One of my favorite thought experiments, famously embodied in many science-fiction stories and in the *Matrix* movies, is to consider intelligent, self-aware beings who are oblivious to the physical world that contains them. For purposes of argument, let's assume that the proponents of strong artificial intelligence have it right, so that such beings could exist. (Given the rapid progress of AI and virtual reality, it's not implausible.)

The "sense organs" of these hypothetical beings would not be portals to the physical world. Their input, instead, would be electrical signals, generated by computers. Thus, the "external world" experienced by these beings—that is, the data flow they interpret as perception—is, in our thought experiment, actually a long series of signals generated by a computer program. Since that "external world" follows instructions crafted by a programmer, it can obey whatever rules the programmer cares to impose.

In this kind of world, each and every one of our principles can be trashed.

We can, for example, imagine an intelligent, self-aware version of Super Mario, whose sensory universe lies inside that game world. Our self-aware Super Mario lives in a universe governed by laws that depend on where he is—specifically, on which level he's achieved. It is a universe, more generally, whose rules can be upended by unpredictable, hidden surprises that the programmers built in—not only quirky rules, but also so-called Easter eggs, which purposely break the rules.

We could construct a world in which astrology is true—where a character's personality and fate really are determined by the position of the stars and planets when they are born. We could program that in. We could program in different kinds of monsters to spring up suddenly when there's an eclipse of the Sun or the Moon. We could allow the characters to cast magic spells that strike down distant enemies at once, locality be damned. Using random numbers, we could also introduce noise, to make the rules unpredictable and imprecise. Computer game designers revel in such possibilities.

We can envision worlds wherein miracles can and do happen. We can envision worlds whose history reaches a preordained climax, according to a planned script. Those thought worlds embody the central ideas of intelligent design theory.

In this way, we've envisioned thought worlds wherein our first principle is misleading and the other principles are flat wrong. These thought experiments demonstrate that those principles are not necessarily true, let alone obvious. The fact that the physical world we presently inhabit appears to obey them is an astonishing discovery. It was not an easy discovery to make—and it is not an easy one to accept.

Anytime I decide to raise my hand, something that contradicts the principles seems to be happening. Indeed, the grammar of the sentence "I decide to raise my hand" says it all:

There is something called "I"—a spirit, or a will—that dictates how a piece of the physical world behaves. It's an illusion, or at least a take on things, that's hard to abandon. But our principles ask us to think differently.

PROPERTIES: WHAT IS MATTER?

> By convention sweet is sweet, bitter is bitter, hot is hot, cold is cold, color is color; but in truth there are only atoms and the void.
>
> —*Democritus, fragment (c. 400 BC)*

That fragment from Democritus can be taken as the founding document of atomism. The second part of the fragment, following the semicolon—"in truth there are only atoms and the void"—is essentially Feynman's "everything is made of atoms."

Democritus's declaration is deeply challenging. It denies the objective reality of the experiences—taste, warmth, color—through which we access the physical world most directly. No doubt what he intended is that we can understand physical reality in terms of basic units—atoms for him, elementary particles for us—that are not themselves sweet, bitter, hot, cold, or colorful. Those perceptions, he suggests, are a highly processed packaging and summary of what's happening under the hood, which is just elementary particles going about their business. But in telling us what properties elementary particles don't have, or at least might not have, Democritus sets up a big, beautiful question: What properties *do* they have?

Democritus's own answer to that question, it appears, was

this: Elementary particles have shape and motion, but no other properties. His elementary particles were rigid bodies, with hooks. The hooks explained how they could stick together to make solids, or different sorts of materials in general. He postulated that his elementary particles have spontaneous motion, or "swerve," as well as preferred positions. The resulting tension between restlessness and desire, according to Democritus, keeps the world a lively place. (Since we have only a few fragments and early commentaries to go on, it's impossible to know exactly what he had in mind. But I think that gets the gist.)

Modern science gives an answer that, while completely different in its details, is no less bold. It is even more radical in its simplicity. Most important, it is backed up by mountains of experimental evidence. According to our present best understanding, the primary properties of matter, from which all its other properties can be derived, are these three:

Mass

Charge

Spin

That's it.

From a philosophical perspective, the key takeaways are that there are very few primary properties, and that they are things you can define and measure precisely. And also this: As Democritus anticipated, the connection of the primary properties—the deep structure of reality—to the everyday appearance of things is quite remote. While it seems to me too strong to say that sweet, bitter, hot, cold, and color are "conventions," it is surely true that it takes quite some doing to trace those things—and the world of everyday experience more generally—to their origins in mass, charge, and spin.

A detailed discussion of mass and charge, including both electric and color charge, can be found in the appendix. Here I will say a bit about spin, which may be the least familiar property.

If you've ever played with a gyroscope, you'll have a head start on understanding the spin of elementary particles. The basic idea of spin is that elementary particles are ideal, frictionless gyroscopes, which never run down.

The fun of a gyroscope, or gyro, is that it moves in ways that are unfamiliar in everyday (nongyro) experience. Specifically, a rapidly spinning gyro resists attempts to alter its axis of rotation. Unless you exert a large force, the orientation of that axis won't change much. We say that the gyro has orientational inertia. That effect is used to guide aircraft and spacecraft, which carry gyros inside to help keep themselves oriented.

The faster a gyro rotates, the more effectively it will resist attempts to change its orientation. By comparing the force with the response, you can define a quantity that measures orientational inertia. It is called angular momentum. Big gyros that rotate rapidly have large angular momentum, and show small responses to applied forces.

Elementary particles, on the other hand, are tiny gyros, indeed. Their angular momentum is *very* small. When angular momentum gets as small as it does for elementary particles, we enter the domain of quantum physics. Quantum mechanics often reveals that quantities which were once thought to be continuously variable actually come in small discrete units, or *quanta*. (This is how quantum mechanics got its name.) So it is for angular momentum. According to quantum mechanics, there is a theoretical minimum to the amount of angular momentum any object can carry. All possible angular momenta are whole-number multiples of that minimal unit.

It turns out that electrons, quarks, and several other kinds of elementary particles carry exactly the theoretical minimum unit of angular momentum. Physicists express that fact by saying that electrons, and the other examples, are particles with spin ½. (There's an interesting mathematical reason why physicists call the basic unit of angular momentum spin ½, rather than spin 1, but it is beyond the scope of this book.)

Before concluding this little introduction to spin, I'd like to add a personal note. Spin changed my life. I always liked math and puzzles, and as a child I loved to play with tops. I majored in mathematics as an undergraduate. During my last semester at the University of Chicago, life on campus got disrupted by student protests. Classes became improvised and semi-voluntary. Peter Freund, a famous physics professor, offered an advanced course on the application of mathematical symmetry to physics. I took the opportunity to sit in on it, even though I wasn't properly prepared.

Professor Freund showed us how some extremely beautiful mathematics, building on the idea of symmetry, leads directly to concrete predictions about observable physical behavior. His enthusiasm, bordering on rapture, shone through his widened eyes as he spoke. To me, the most impressive example of this connection was—and still is—the quantum theory of angular momentum, which he showed us. When a spinning particle decays into several other spinning particles (which is a very common situation in the quantum world), the quantum theory of angular momentum makes predictions about relationships among the directions in which decay products emerge and the orientations of their rotation axes. Working out those predictions requires substantial calculations, and the behaviors they predict are anything but obvious. Amazingly, though, they work.

To experience the deep harmony between two different universes—the universe of beautiful ideas and the universe of physical behavior—was for me a kind of spiritual awakening. It became my vocation. I haven't been disappointed.

Philosophy of Properties

Let me emphasize, again, that the most important and remarkable point about our trinity of properties—mass, charge, and spin—is simply that there are so few of them. For any elementary particle, once you've specified the magnitude of those three things, together with its position and velocity, you've described it completely.

How different it is for the objects of everyday life! Objects we commonly encounter have all kinds of properties: sizes, shapes, colors, smells, tastes, and many others. And when we describe a person, it is useful to specify their gender, age, personality, state of mind, and a host of other variables. All those properties of objects or people supply more or less independent pieces of information about them. No subset determines the rest. Evidently, there is a startling contrast between the stark simplicity of the basic ingredients and the complexity of the products they produce, just as Democritus suspected.

Contrary to Democritus, though, our modern basic ingredients don't have hooks. They aren't even solid bodies. Indeed, though it's convenient to call them "elementary particles," they aren't really particles. (That is, they have little in common with what the word "particle" suggests.) Our modern fundamental ingredients have no intrinsic size or shape. If we insist on visualizing them, we should think of structureless points where concentrations of mass, charge, and spin reside. We have, in place of "atoms and the void," space-time and properties.

THE PARTICULARS

Not all elementary particles are created equal. They play different roles in our understanding of the world. A few dominate everyday life. A few more come into their own in astronomy and astrophysics. And then there are others whose role in the big scheme of things is not entirely clear.

In other words, we have particles of construction, particles of change, and bonus particles. They are all fascinating to professional physicists and astronomers, but the particles of construction are by far the most important for understanding the world we experience, and I'll focus on them here. Some further discussion of the others appears in the appendix.

Particles of Construction

Roughly defined, "ordinary matter" is the sort of matter we're made of and that we commonly encounter in biology, chemistry, geology, and engineering. It is a major achievement of modern science that we can also define ordinary matter in a quite different way, and more precisely: It is the matter we can build up from electrons; photons; two kinds of quarks, commonly called "up" and "down" quarks; and gluons.

Thus, we can construct the matter that we encounter in ordinary life, and that our bodies are built from, using exactly five kinds of elementary particles as ingredients, each precisely defined by a few limpid properties.

Here is a table that lists those particles and their properties:

	mass	**electric charge**	**color charge**	**spin**
electron	1	-1	no	½
photon	0	0	no	1
u quark	10*	⅔	yes	½
d quark	20*	-⅓	yes	½
gluon	0	0	yes	1

(The asterisks will be explained in due course.)

To kick-start this census, let me quickly recall the "classic" description of atoms, coming out of the early twentieth century, which we'll be refining. In that description, an atom consists of a small central nucleus surrounded by a cloud of electrons. Electrical attraction binds the electrons to the nucleus. The nucleus contains almost all the mass of the atom, and all of its positive electric charge.

The nucleus in turn is formed out of protons and neutrons. Both protons and neutrons weigh about two thousand times the mass of electrons. Protons carry positive electric charge, such that the positive electric charge of one proton exactly balances one electron's negative charge. Neutrons carry zero electric charge. Thus, when the number of electrons surrounding a nucleus is equal to the number of protons within it, the atom as a whole carries zero electric charge, and is electrically neutral.

Electrons were the first elementary particles to be discovered, and in many ways they are the most important. Electrons were first clearly identified by J. J. Thomson in 1897. He studied electrical discharges—essentially, artificial lightning—in highly evacuated "vacuum" tubes. The tubes weren't quite empty inside—otherwise there would be no electrons to study—but they were empty enough to allow the particles within them some running room. (Today, we understand that when you apply very strong electric fields—in other words, high voltages—across highly evacuated tubes, you "ionize" the atoms, stripping off electrons. The charged particles move in response to the applied fields, and set off some sparks as they do.) By applying electric and magnetic fields and looking at how much different parts of the discharges bent, Thomson identified an especially meaningful component of the discharges. This special component appears in all discharges—that is, no matter what gas you fill the tube with—and the way it bends in response to magnetic fields is especially simple. In fact, the shape of this responsive "lightning bolt" matches the path you calculate, using the laws of electricity and magnetism, for the motion of charged, massive points, with specific values of the charge and mass. Naturally, Thomson proposed that his special discharges were made up of particles which have that much mass and carry that much charge. This was the birth of electrons. The observation that electron streams appeared in all discharges, whatever the starting gas, suggested that they were a basic, universal building block of matter.

Thomson's pioneering work inspired many follow-up investigations. Before long, those deep dives into the nature of matter gave birth to a technology that is now both familiar and ubiquitous—electronics. Its importance would be hard to overstate.

The behavior of electrons has been studied from many angles, in many different kinds of experiments. For example, as I mentioned earlier, people have measured the tiny magnetic fields that spinning electrons—that is, all electrons—generate. The magnitude of those fields can be predicted, by calculation, based on the hypothesis that electrons have mass, electric charge, spin, and no other properties. The predictions can be calculated to very high accuracy, and the magnetic fields can be measured to very high accuracy—each at the level of parts per billion. Happily, they agree.

Accurate agreement between the predicted behavior of an ideally simple model of electrons and experimental observations is what we mean, operationally, when we say that electrons are elementary particles. If electrons, like atoms, had significant internal structure, then they wouldn't behave so simply. If, for example, an electron's electric charge were uniformly distributed in a little ball, rather than concentrated at a point, then the predicted value of the electron's magnetic field would be different, and it would no longer agree with what people have measured. (Of course, if the ball were small enough, the difference might not be noticeable. What we can say for sure is that Nature hasn't encouraged us to bring in that complication.)

The same kind of justification could be offered for each of the elementary particles we're going to discuss. They've earned the title "elementary," until proven otherwise, because that stringent assumption—that they have a very few properties and no others—has lots of impressively successful consequences.

In the table of elementary particles and their properties, I've used the electron mass to set the scale for all other elementary particles' masses, so by definition it is 1. I've also used, as is conventional, the electron's electric charge as the standard of electric charge. But here there's a slight complication, courtesy of a great personal hero of mine, Benjamin Franklin. Before he became known as a statesman and diplomat, Franklin made pioneering contributions to early electrical science. He discovered the conservation of electrical charge, and also proved that it comes in both positive and negative varieties.

By being first, Franklin got to choose which kind of charge to call positive and which negative. He chose to call the charge that accumulates on glass, after it is rubbed with silk, positive. This was long before people knew about electrons. Unfortunately, it turns out that according to Franklin's choice the electron's charge is negative. It's much too late to undo that choice, since it has seeped into thousands of books, papers, and circuit diagrams. Therefore, we list the electron's electric charge as –1.

Photons were the next elementary particles to be discovered. The existence of light was a "discovery" known throughout the animal world, and arguably to plants, long before human history began. The discovery that light comes in discrete units, on the other hand, started as a theoretical proposal. Photons are the elementary units of light.

Einstein first made this suggestion during his "miracle year" of 1905—the same year that contained special relativity, the existence of atoms (Brownian motion), and $E = mc^2$. He called it the hypothesis of light quanta. (The word "photon" was introduced later, in 1925, by the prominent chemist Gilbert Lewis.) It was a revolutionary proposal, which opened to bad reviews. Eight years later, in 1913, toward the end of his glowing recommendation of Einstein for membership in the Prussian Academy of Sciences, Max Planck apologized for Einstein's embarrassing absurdity by writing, "That sometimes, as for instance in his hypothesis on light quanta, he may have gone overboard in his speculations should not be held against him."

Ironically, Einstein's proposal was based on Planck's work. Planck had argued that light was emitted and absorbed in lumps based on experiments measuring the glow from heated bodies (so-called blackbody radiation). Einstein interpreted this as evidence that light was made of lumps, period. He used his more specific interpretation to make predictions about several other kinds of possible experiments. The proposed new experiments were very challenging for 1905 technology. It was only in 1914—one year after Planck's letter—that Robert Millikan carried out truly decisive tests of Einstein's proposal.

Though he surely deserved several others, Einstein received his only Nobel Prize in 1921, for his work on light quanta. Einstein himself regarded this as his most revolutionary work.

When you study the behavior of matter at higher energies than was possible in the early twentieth century, you come upon individual photons that carry significant energy and momentum. This makes them much easier to identify as particles. High-energy photons are known as gamma rays. You can use a Geiger counter to hear gamma rays announcing their arrival, click by click.

We should consider photons, together with electrons and atomic nuclei, as components of atoms. Indeed, photons are the original "gluons." It is photons, in their collective incarnation as electric fields, that glue atoms together, binding electrons to their nuclei.

Protons and neutrons are not elementary particles. Their behavior proves to be too complicated for that description to be viable. The model of protons and neutrons we use today is easy to describe, though it was not easy to discover or to prove. It runs broadly parallel to the theory of atoms. Two kinds of electron-like particles—called *u* quarks and *d* quarks—get bound together by photon-like particles called gluons.

Though the basic idea is similar, there are some notable differences between how atoms are assembled (from electrons, photons, and a nucleus) and how protons are assembled (from quarks and gluons):

- Strong forces, which are controlled by color charge, are much stronger than electromagnetic forces, which are controlled by electric charge. This is why atomic nuclei, which are bound together tightly by the strong force, are much smaller than atoms.
- While electrons always repel one another, quarks, because their color charges come in three varieties, feel more complex forces, which can be attractive. This possibility allows quarks, in contrast to electrons, to bind together without requiring a "nucleus" made of something else.
- While photons are electrically neutral—that is, they have zero electric charge—their strong force analogues, the color gluons, are not color charge neutral. Gluons feel the strong force, just as much as (in fact, more than) quarks do. This is another reason why protons and neutrons are more homogeneous than atoms: The carriers of the force are also under its influence.

To complete our account of quarks and gluons, we need to discuss their masses.* For gluons this is simple: Like photons, gluons have zero mass. For quarks, the most important thing to note is that while their mass is large relative to electrons, it is very small relative to protons or neutrons.

* Quarks also carry nonvanishing electric charges. Here there is a distinction between two kinds of quarks—the *u* quark, whose electric charge is $\frac{2}{3}$, and the *d* quark, whose electric charge is $-\frac{1}{3}$. Protons coalesce around two *u* quarks and one *d* quark, so their electric charge is $\frac{2}{3} + \frac{2}{3} - \frac{1}{3} = 1$. Neutrons coalesce around one *u* quark and two *d* quarks, so their electric charge is $\frac{2}{3} - \frac{1}{3} - \frac{1}{3} = 0$.

It might seem paradoxical that the mass of protons is much larger than the total mass of the things they're made of. In truth it points to a crowning achievement in the human understanding of Nature: understanding the origin of our mass, in energy. We'll discuss it further in the next chapter.

It is difficult to measure the masses of *u* quarks and *d* quarks accurately, because it is difficult to discern the influence of those masses amid other, larger effects. That is why I've put asterisks in the table next to the best estimates of their values.

We should add the graviton to our list of particles of construction. The graviton is the particle from which gravitational fields are made. Photons bind together atoms and molecules; gluons bind together quarks, protons, and atomic nuclei; gravitons bind planets, stars, galaxies, and big things in general.

	mass	**electric charge**	**color charge**	**spin**
graviton	0	0	no	2

Gravitons have never been observed as individual particles, because their interactions with ordinary matter are far too feeble for that to be practical. What has been observed are gravitational forces—and, recently, gravitational waves. Theoretically, those observable effects arise from the cumulative action of many individual gravitons.

Each of the properties of gravitons I've listed has a clear connection to observed features of the force which gravitons generate—that is, gravity. Since gravitons have zero electric charge and no color charge, individually they interact only feebly with ordinary matter. Yet because they have zero mass, gravitons can be made cheaply in great numbers, to generate gravitational fields and gravitational waves.

Their relatively large spin implies that gravitons' interactions are more intricate than those of other elementary particles. Indeed, one can show that the main features of Einstein's theory of gravity, general relativity, follow directly from those spin-derived properties of gravitons. The fact that you can do so is an impressive demonstration of the power of our three primary properties of matter—mass, charge, and spin—to account for matter's behavior fully. Einstein himself originally arrived at general relativity by an incredibly brilliant but much less straightforward path.

This concludes our tour of the particles of construction. If this is your first encounter with these ideas, the unfamiliarity of the concepts and their embodiments might be a little dizzying. The fundamental message, though, should shine through: The physical world is constructed using very few kinds of ingredients. Moreover, those ingredients are ideally simple, in the sense that they have only a handful of properties.

THE FUTURE OF INGREDIENTS

The list of elementary particles is significantly shorter than the English alphabet, and much shorter than Mendeleev's periodic table of chemical elements. Taken together with the laws describing forces—four, to be exact—this list of ingredients gives us a powerful, successful description of matter. We'll be exploring all that in the next chapter. There we'll also discuss tantalizing hints and ideas about how we might get an even more compact description.

But before we get to that, I want to consider the future of world-building ingredients from a different, more practical angle. I'll describe two promising strategies for making useful new "elementary particles." Both strategies are inspired by Nature. One strategy, inspired by physics, works from the outside in. The other, inspired by biology, works from the inside out.

Designer Particles, Take 1: Brave New Worlds

We can think about materials using the same ideas we use to analyze the world as a whole. When you inject a bit of energy into a material, or a bit of electric charge or spin, the resulting disturbance will generally cohere into a few lumps, or quanta. These "otherworldly" lumps, called quasiparticles, can have quite different properties from the elementary particles we encounter in empty space.

Holes are a simple but extremely important class of quasiparticles. Inside a typical solid there are many electrons. When the solid is undisturbed, in equilibrium, the electrons arrange themselves in a definite pattern. Now imagine plucking one out. The resulting state will have an empty spot where an electron "ought to be." After things settle down, which can happen quite quickly, what's usually left behind is a quasiparticle, which, since it arose from the absence of an electron, carries electric charge +1 (here we recall that the electron's charge is –1). We call it a hole.

Holes give us positively charged (quasi)particles that are much lighter and easier to manipulate than their closest empty-space analogues, protons. Holes are star players in transistors, and in modern electronics more generally. Understanding how to make and use holes changed the world.

In other cases, quasiparticles descend directly from the elementary particles of empty space, but when they are inside the material they acquire distinctly different properties than they had in empty space. An elegant example of this occurs in superconductivity. When photons enter a superconductor, their mass changes from zero to a tiny, but nonzero, value. (The value varies depending on the superconductor in question; a millionth of the electron's mass is typical.) Indeed, to sophisticated physicists the fact that photons acquire mass is the essence of superconductivity.

My earliest research in physics focused on elementary particles in the traditional sense. But long before that, during a school trip to Bell Labs, I had an experience that stuck in my mind, and eventually changed my life. During our visit, we listened to a talk in which one of the scientists, trying to explain his work to us, mentioned that phonons are the quanta of vibration. I didn't understand what he was talking about, but I thought it was the coolest thing I'd ever heard—three weird concepts, each with a resonant name, somehow wrapped into one. On the way home, puzzling it out, I managed to convince myself that his message was that materials are like worlds in themselves, different from ours, which are homes to their own kinds of particles. I loved that idea.

It's slow work to invent new kinds of elementary particles.

All of the elementary particles that I discussed above, and also those in the appendix, were either known or confidently anticipated already in the 1970s. On the other hand, there's enormous scope for imagination and creativity in the worlds of quasiparticles. That school expedition, in retrospect, was a glimpse of new horizons.

Fifteen years later, I finally reached those horizons. Here I'll just mention one highlight. *Anyons* are quasiparticles that have a simple kind of memory. I introduced them, and gave them their name, in 1982. At first, it was purely an intellectual exercise. I wanted to demonstrate that quasiparticles could support a tiny memory, as an additional property. (Later I found out that two Norwegian physicists, Jon Magne Leinaas and Jan Myrheim, had discussed related ideas earlier.) At that point, I didn't have any particular material in mind.

A few months later, though, I learned about a discovery called the fractional quantum Hall effect (FQHE).* Within FQHE materials, an injected electron divides into several quasiparticles, each of which carries a fraction of its electric charge. I realized that those quasiparticles must exert very peculiar forces on one another, which made me suspect they might be anyons. In 1984, working with Dan Arovas and J. Robert Schrieffer, I managed to prove it.

Since then I've had a lot of fun with anyons, and hundreds of other physicists have joined the party. People hope to use anyons as building blocks for quantum computers, because you can use their memory to store and manipulate information.

* Robert Laughlin, Horst Störmer, and Dan Tsui shared the 1998 Nobel Prize in Physics for this discovery.

Microsoft has made a big investment in research toward that goal.

Physicists and creative engineers have proposed many other interesting and potentially useful new kinds of quasiparticles. They have endearing names like spinon, plasmon, polariton, fluxon, and my favorite: exciton. Some are good at capturing radiant energy, while others are good at transporting energy from one place to another. Those two talents can be combined to design efficient solar energy systems.

Brave new material worlds with wondrous quasiparticles will be an important part of the future of matter. The burgeoning field of metamaterials designs them systematically.

Once you get to thinking about materials as homes to quasiparticles, a profound question is not far off: Can we consider "empty space" itself to be a material, whose quasiparticles are our "elementary particles"? We can, and we should. It is a very fruitful line of thought, as you'll see in later chapters.

Designer Particles, Take 2: Smart Materials

Biology suggests another direction for the future of matter. Cells are the "elementary particles" of advanced life forms. They come in many shapes and sizes, but they share a large bag of tricks that enable them to function as repositories of information and as chemical factories. They also have sophisticated interfaces with the external world, which enable them to gather resources and exchange information.

Biological cells are far from being simple physical objects. It is a daunting challenge to construct from scratch artificial units that have those core functionalities of cells. If one could, then the door would be open to making new cell-like units that could fill in for diseased or senescent cells, or to bring in new capabilities like digesting toxic waste into harmless or useful materials. A more practical short-term strategy, used now with increasing success by many molecular biologists, is to tweak existing cell types.

On the other hand, it is possible to be inspired by biology without being literal about it. Cars aren't souped-up horses, nor are airplanes souped-up birds, nor do useful robots have to resemble humans. The most unique feature of biological cells, to which present-day human engineering has no close analogue, is the power of modulated self-reproduction. In appropriate, reasonably forgiving environments, cells will gather ingredients to make new cells that are close, but not necessarily exact, copies of themselves. The differences are not random, but follow programs contained in the cell itself.

Self-reproduction unleashes the power of exponential growth. Starting with one cell, after ten generations of doubling one has more than a thousand cells, and after forty or so generations one has trillions of cells, which are enough to make a human body. Programmed differences—that is, modulations—can (and do) generate specialized cells appropriate to different functions, as is the case with muscle cells, blood cells, and neurons.

It should be possible to realize the powerful strategy of modulated self-reproduction in artificial units that are considerably simpler than biological cells, especially if their intended use is less complex and delicate than producing a viable biological organism. Some grand projects, such as terraforming planets or constructing mountain-sized computers, whose realization is both highly repetitive in structure and forgiving in detail, are plausibly of this kind. Modulated self-reproduction is such a powerful concept that I am confident it will feature prominently in the engineering of the future.

4

THERE ARE VERY FEW LAWS

The way that the fundamental* physical laws work is quite different from how human laws work. There are many human laws, and they differ from place to place and change over time. Human laws presuppose that there are different options for behavior, and propose reactions to them. Human laws do not support long chains of reasoning that lead to unambiguous conclusions, and experts often differ about their meaning.

Fundamental physical laws differ from human laws on each of those counts. There are very few of them, and they are the same everywhere and always. Physical laws simply describe what will happen. Physical laws are expressed as mathematical equations among precisely defined quantities, leaving no room for ambiguity or disagreement among competent experts.

---

* Here by "fundamental" laws I mean laws that cannot be derived, even in principle, from other laws. Laws can be profoundly important and central to our understanding of nature without being "fundamental" in this sense. The second law of thermodynamics is a good example of that.

Drawing out their consequences is merely a matter of calculation. You can program a computer to do it.

A child's conception of how the world works, which most people carry into adulthood by default, stands much closer to the model of human law than to the ideal of physical law. We have the direct experience of weighing options and making choices. Our mental choices seem to make a difference in the physical world. Specifically, they seem to control how our bodies move. We form expectations for how people and things will behave based on rules of thumb, and only rarely through chains of logic and calculation. Nobody walks, rides a bicycle, or catches a fly ball by working up from Newton's laws of motion, let alone the quantum theory of matter.

To reach fundamental understanding, we need to rethink those experiences and childlike methods. Only then can we graduate from human law to physical law.

THE TRIUMPH OF LOCALITY AND THE GLORY OF FIELDS

Newton's *Principia*, published in 1687, established a powerful framework for understanding the physical world that dominated science well into the nineteenth century. Within this framework, laws express how bodies exert forces upon one another. The model of a successful law was Newton's law of gravity. According to that law, bodies attract one another with a force that is proportional to the product of their masses and decreases as the square of the distance between them.

When people began to grapple with other kinds of forces—electric and magnetic forces, to be specific—they tried to use the same basic framework. Early results were encouraging. Coulomb's law for electric forces, for example, echoes Newton's law for gravitational forces, with electric charge taking the place of mass.

But it didn't work as neatly for magnetism. Magnetic forces turned out to depend on velocity, as well as on position, in a complicated way. Then, when people studied situations where both electricity and magnetism operated at the same time, the complications multiplied.

Michael Faraday (1791–1867), a self-educated experimental genius of humble origins, could not follow the intricate mathematics of these complicated force-laws. He thought for himself, instead, in imagery. He visualized that electrically and magnetically active bodies extend influence through space, as a sort of aura or atmosphere, even where no other bodies are around to feel that influence. Today, we call these activations of space electric and magnetic fields. Faraday used more vivid language; he called them "lines of force." As James Clerk Maxwell (1831–1879), the spectacularly gifted theorist who became Faraday's disciple and evangelist, put it, "Faraday, in his mind's eye, saw lines of force traversing all space where the mathematicians saw centres of force attracting at a distance: Faraday saw a medium where they saw nothing but distance: Faraday sought the seat of the phenomena in real actions going on in the medium."

Guided by his unorthodox ideas, Faraday soon discovered a remarkable effect that was difficult even to state without referring to his fields. This is his law of induction, according to which magnetic fields that change in time produce circulating electric fields. With that discovery, he revealed that fields have a life of their own.

An everyday experience with water provides a familiar model to illustrate how a space-filling medium can generate forces between distant bodies, through local action, as Faraday envisaged. If a moving boat, or jet ski, creates a disturbance in a lake, the influence of that disturbance spreads gradually through the lake, as moving water at one location pushes water nearby—and *only* water nearby. And so eventually, even if they're far from the source, swimmers in the lake will feel a force when the wave arrives. I've had that annoying experience many times. It would be worse if it came without warning, but usually I see the wave coming. Locality is a blessing—it means you can't be taken completely by surprise.

Faraday's more complete vision of locality inspired a revolution in physics. Since electromagnetic fields, which fill space, have a life of their own, they must be included among the world's ingredients. The Newtonian framework, based on particles in space—harkening back to Democritus's "atoms and the void"—wouldn't cut it anymore. Thus, our description of the world was profoundly enriched. As Maxwell wrote:

> The vast interplanetary and interstellar regions will no longer be regarded as waste places in the universe, which the Creator has not seen fit to fill with the symbols of the manifold order of His kingdom. We shall find them to be already full of this wonderful medium; so full, that no human power can remove it from the smallest portion

of Space, or produce the slightest flaw in its infinite continuity.

If Maxwell's rapturous prose seems excessive, let's consider how he got there. When Maxwell decided, early in his career, to take up the study of electricity and magnetism, he was galvanized by Faraday's conceptions and discoveries. He resolved to build upon Faraday's intuitive field concept, rather than retreat to the much better-developed and more popular Newtonian framework. Maxwell put forward that

> whenever energy is transmitted from one body to another in time, there must be a medium or substance in which the energy exists after it leaves one body and before it reaches the other. . . . And if we admit this medium as an hypothesis, I think it ought to occupy a prominent place in our investigations, and that we ought to construct a mental representation of all the details of its action.

Upon spelling out this new viewpoint mathematically, Maxwell discovered that in order to get consistent equations he needed to supplement Faraday's law of induction with another one, in which the roles of electric and magnetic fields are reversed. According to Maxwell's law of induction, electric fields that change in time produce circulating magnetic fields.

When he married the two field-based induction laws—Faraday's and his own—Maxwell discovered that they gave birth to a dramatic new effect. One could have a self-restoring, permanent, traveling disturbance in electric and magnetic fields. Changing electric fields induce changing magnetic fields induce changing electric fields induce changing magnetic fields . . . Those disturbances, he calculated, should travel at the speed of light, which had been measured independently. Maxwell immediately proposed, "The agreement of the results seems to show that light and magnetism are affections of the same substance, and that light is an electromagnetic disturbance propagated through the field according to electromagnetic laws."

He was right.

The possible electromagnetic disturbances include visible light—all the wavelengths perceptible to our eyes—and much more. Maxwell predicted the existence of stretched and compressed versions of visible light, including new forms of radiation that were totally unknown and unexpected at the time. Today, we call them radio waves, microwaves, infrared and ultraviolet radiation, x-rays, and gamma rays.

The decisive experimental test of Maxwell's equations came more than twenty years after they were first proposed. To achieve it, Heinrich Hertz designed and built the first radio transmitters and receivers. Hertz's goal was to turn beautiful ideas into physical realities. He felt he had succeeded at that. "One cannot escape the feeling," he wrote, "that these mathematical formulae have an independent existence and an intelligence of their own, that they are wiser than we are, wiser even than their discoverers, that we get more out of them than was originally put into them."

The work of Faraday, Maxwell, and Hertz spanned most of the nineteenth century. It established space-filling fields as a new kind of ingredient in the fundamental description of the world.

Force and Substance: Quantum Fields

At first, fields were considered an additional ingredient in the recipe for the physical world, supplementing particles. Over the twentieth century, fields took over completely. We now understand particles as manifestations of a deeper, fuller reality. Particles are avatars of fields.

As we mentioned earlier, Einstein, building on the work of Planck, proposed that light comes in discrete units, particles that Einstein called light-quanta, and which we now call photons. Einstein's proposal initially got a chilly reception from the physics community, because it seemed difficult to reconcile the idea that light comes in particles with Maxwell's field-based understanding of light. Maxwell's theory had scored many triumphs, including Hertz's epochal discovery, and was reinforced by detailed study of the new forms of radiation it predicted.

Fields, being continuously extended through space, appear to be very different from particles. It was hard to imagine how light could be both, yet experimental facts demanded it.

The different aspects of light—field and particle—get reconciled in the concept of a quantum field. Quantum fields, as their name suggests, are still fields (that is, space-filling media). There are quantum versions of both electric and magnetic fields. They continue to satisfy the same equations—Maxwell's equations—that nineteenth-century physicists proposed for electric and magnetic fields, before anybody knew about quantum mechanics.

But the quantum versions of the electric and magnetic fields satisfy additional equations. The additional equations usually go by the rather forbidding name "commutation relations," but I will use the less formal name "quantum conditions." Whatever you call them, these additional equations express the essence of quantum theory in mathematical form. Werner Heisenberg introduced the general idea of quantum conditions in 1925, when he was twenty-four years old. Paul Dirac introduced the specific quantum conditions that apply to electric and magnetic fields shortly afterward, in 1926. He, too, was twenty-four years old.

With more equations to satisfy, there are fewer solutions. As we've discussed, Maxwell discovered that light is a kind of self-perpetuating, moving excitation among electric and magnetic fields. Not all of his solutions, however, also satisfy the quantum conditions. The allowed solutions must satisfy a specific relationship between their energy and their frequency (that is, the rate at which the fields oscillate). I will state that important relationship both in words and, alternatively, as a simple equation. The relationship is that the energy of the excitation must be equal to a nonzero constant, called Planck's constant, multiplied by the frequency. As an equation, it reads $E = h\nu$, where E is the energy, ν is the frequency, and h is Planck's constant. This relationship, not coincidentally, is the one that Planck proposed in 1900 and that Einstein seized on in 1905, to predict the existence of photons. It is called the Planck-Einstein formula. It took twenty years to digest their revolutionary proposal, closely based on experimental results, before physicists reached a consistent theoretical interpretation, as presented here. We have both Maxwell's equations *and* discrete units of light.

This grand story of electromagnetic fields and photons leads directly to another key insight. It explains why, and how, Nature produces vast numbers of interchangeable parts.

If our account of the fundamental ingredients had ended at the level of elementary particles, it would leave a basic question unanswered. For at that level, we must postulate that each kind of elementary particle exists in many identical copies: many identical photons, many identical electrons, and so forth.

In the history of human manufacturing, the introduction of standardized, interchangeable parts was a great innovation. To achieve it, new kinds of machines and materials had to be invented so that accurate templates could be made and maintained. And even so, the parts, once made, are subject to wear and tear, and eventually cease to be identical.

Photons, on the other hand, are observed to have the same properties, whenever and wherever they are found. The light of a given color is the same thing—it has the same properties, and interacts with matter in the same way—whatever its source. Likewise, electrons are precisely the same wherever they are found. If the electrons within different atoms of carbon, for example, did not have identical properties, then each carbon atom would have different properties, and chemistry would not work.

How does Nature do it? By tracing the common origin of all photons to a common, universal electromagnetic field, we come to understand their otherwise baffling sameness. And we are led, by analogy, to introduce a field—call it the electron field—whose excitations are electrons. All electrons have the same properties, because each one is an excitation in the same universal field.

Fields are necessary to achieve locality, and quantum fields produce particles. Following this chain of logic, we obtain a deeper understanding of why particles exist, and of their amazing interchangeability. There is no need to introduce two different sorts of fundamental ingredients, fields and particles, after all. Fields rule. Quantum fields, that is.

Going back to the origin of the field concept, in Faraday's attempts to picture electric and magnetic influences in space, we can recognize another way that quantum fields unify our picture of the world. The same quantum electric and magnetic fields that produce photons also produce, according to Faraday's visions—and Maxwell's equations—electric and magnetic forces.

To sum up:

> From forces we are led to fields, and from (quantum) fields, we are led to particles.
>
> From particles we are led to (quantum) fields, and from fields, we are led to forces.
>
> Thus, we come to understand that substance and force are two aspects of a common underlying reality.

FOUR FORCES

In this section, I will briefly sketch our best understanding of the nature of the four known forces, using the framework we discussed in the preceding chapter: principles and properties, embodied in a few kinds of particles. One layer deeper, the particles get replaced by fields, as we just discussed.

The four forces are:

- electromagnetism, or in its full quantum glory, quantum electrodynamics (QED)
- the strong force, or in its full quantum glory, quantum chromodynamics (QCD)
- gravity, as captured in Einstein's general relativity
- the weak force

The electromagnetic and strong forces dominate our understanding of terrestrial matter. The electromagnetic force holds atoms together and governs their structure. It also describes how they interact with light. The strong force holds atomic nuclei together and governs their structure.

Gravity, as it acts between elementary particles, is very feeble. But when many particles are involved, its influence accumulates, and it comes to dominate interactions between large bodies.

The weak force governs processes of transformation. It causes some otherwise stable particles to decay, as in some forms of radioactivity. Notably, too, it mediates energy-releasing interactions that power stars, including our Sun.

Before we plunge into more detail, I'd like to explain two choices I've made. The first is simply a choice of words. Physicists often speak of the four "interactions," as opposed to the four "forces." There is a legitimate argument for that choice. "Force" has a precise technical meaning in Newtonian mechanics, where it denotes a potential cause of motion. But in the phrase "weak force," for example, the same word must be understood differently, to include interactions that do other things (namely, processes that change one kind of particle into another). Nevertheless, I'll stick to "weak force," since it is less stilted* than "weak interaction."

The second choice I've made gets to the heart of what I hope to accomplish in this book. The crowning glory of our theories of the four forces is that they can be expressed precisely and accurately in a few mathematical equations. This means something concrete, philosophically, which you do not need mathematical training to understand. It means that it is possible to translate the theories, without loss of content, into reasonably short computer programs. You could, of course, then combine the four programs for the separate forces into one master program. The master program—the operating system of the physical world—would still be *much* shorter than, say, the operating system that runs your computer.

But the flip side of that extraordinary "data compression" is that its information is encoded in something very different

* That is, more forceful.

from any natural human language. The raw equations, or their equivalent in a computer program, use symbols and concepts that are quite remote from the everyday experiences that natural language builds on. It takes a lot of calculation and interpretation to get from the raw equations to consequences that are easy to talk about. So here I had to make a choice—a whole series of choices, actually—about how raw to get and which consequences to emphasize. The overarching message remains—that a very few laws suffice to govern the physical world.

QUANTUM ELECTRODYNAMICS (QED)

The Electric Atom

The basic rules for electromagnetic interactions, starting with Coulomb's law for electric forces and culminating in Maxwell's equations, were deduced from experiments with human-sized objects. Nevertheless, as people began to explore the subatomic world, they assumed, by default, that the only important forces in atomic physics are electromagnetic forces, and that they could continue to use Maxwell's equations to describe those forces. It was the radically conservative thing to do.

That bold strategy works amazingly well. If you accept the basic picture that most of the mass of an atom, and all of its positive electric charge, is concentrated in a small nucleus, and that the remainder consists of electrons, then Maxwell's equations plus a quantum condition—this time, for the electron field—do the rest. Together they give us a model of atoms that is both precise and rich in consequences.

How do we know it's right? *Atoms sing songs that bare their souls, in light.* Allowing for a little poetic license, this phrase describes the art and science of spectroscopy.

Spectroscopy

Let's begin at the beginning, with the photon* and electron fields. The photon field gave us, through its quantum condition, photons. Photons, being electrically neutral, do not influence one another directly.

The electron field gives us, through its quantum condition, electrons. Electrons do influence one another, through electric forces. Because of that, we can't build up all the excitations of the electron field simply by adding up the most basic ones independently. But when electrons are reasonably far apart, the energy involved in their interactions is much less than the energy tied up in their mass (that is, $E = mc^2$), so they retain their integrity. In other words, the basic excitations of the electron field look like a bunch of little particles—electrons—that influence each other. That field-based ferment provides the usual starting point both for elementary science courses and for advanced chemistry and biology texts.

To model an atom, we introduce the influence of a nucleus, and let it act among excitations of the electron field that contain enough electrons to balance the positive electric charge of

* The terms "photon field" and "electromagnetic field" are interchangeable.

that nucleus. Within that setup, the accurate equations for the electron field can get quite complicated, because we need to include both the influence of the nucleus on the electrons and the influence of the electrons on one another. This is the beginning of the long but inexhaustible story of atomic physics and chemistry, based on fundamentals. Many talented people spend their entire careers exploring parts of it.

Our goal here, however, is both broader and more limited. We want to understand in a very general way what some of the most basic predictions of atomic physics look like, and how they connect to fundamentals. For that purpose, the central result of atomic physics is beautifully simple to state: *By studying the colors of light that atoms emit, we can collect rich and detailed information about how they work.*

Here's how that goes: An atom can exist in states with different total energy. The allowed energies form, because of the quantum condition, a pattern of discrete values. States with higher energy can decay into states with lower energy, by radiating a photon. The photon's energy reads out the difference in energy between the initial and final atomic state energies. As Planck and Einstein taught us, the energy of a photon is related to its frequency—or, equivalently, its color. And that is something that is practical to measure.

The array of colors that an atom emits is called its spectrum. The study of spectra is called spectroscopy. Spectroscopy is among the most powerful tools we have to communicate with Nature. It can be used to study not only electrically neutral atoms, but molecules, too, or atoms that are not electrically neutral (ions), or anything else that emits photons.

In 1913, before quantum mechanics assumed its modern, mature form, Niels Bohr invented some rules to restrict the possible energies of hydrogen atoms. Bohr pulled his rules out of thin air, using inspired guesswork. They predicted a spectrum that agreed remarkably well with existing observations. This was not entirely surprising, since they were devised with those observations in mind. More impressive was that Bohr's framework led to additional predictions, which all worked. When Einstein, attending a seminar, first learned of one notable confirmation, he was visibly moved, and said (referring to Bohr's work), "Then it is one of the greatest discoveries."

Bohr's swashbuckling success was enormously influential. It inspired people to look for more general, logically coherent quantum conditions. Today, we see Bohr's rules, together with the Planck-Einstein relation, as the precursors of our modern quantum conditions.

Einstein called Bohr's work "the highest form of musicality in the sphere of thought." Yet modern quantum mechanics, its descendant, is far more harmonious—and the resemblance of its equations to the equations that arise in music is uncanny.

The equations for the electron field around a nucleus, specifically, resemble the equations for a gong constructed from a strange material. Within that metaphor, the spectrum of colors of light emitted by the atom corresponds to the spectrum of tones emitted by the gong. Both reflect their instruments' stable patterns of vibration. But the spectra of atoms are not designed for musical purposes. They do not form the notes of any sensible scale. Especially when more than one electron is involved, the allowed patterns of vibration can become very intricate. Atomic spectra are perfectly definite, and in principle they can be calculated, but they are complicated.

The disciplined complexity of spectra is a gift to human understanding. Since each distinct kind of atom emits a distinct pattern of light, atomic spectra form a kind of signature, or fingerprint. Thus, simply by looking—and paying careful attention to color!—we can discern the identity and study the behavior of atoms that are far removed from us in space and time. The cosmos becomes a giant, well-equipped chemistry lab. For that reason, spectroscopy is a mainstay of astrophysics and cosmology.

Spectroscopy also allows us to test our fundamentals. Since—so far—our accurate theoretical calculations of these spectra, in the cases where we've managed to do them, agree with precise observations, we gain confidence that we've got the laws right. And since—so far—astronomers and chemists have seen the same set of atomic spectra everywhere and at every time they've looked, we conclude that the same laws operate upon the same basic materials everywhere in the universe and throughout its history.

QUANTUM CHROMODYNAMICS (QCD)

The wonderful results of atomic modeling and spectroscopy take off from the bold assumption that atoms have tiny nuclei that contain all of their positive electric charge and almost all of their mass. Following that success, the next item on the fundamental physics agenda, logically, was to understand those nuclei. It launched an exploration that dominated research in physics over much of the twentieth century, and was full of surprising discoveries and twists and turns. Here, so that we can get right to fundamentals, I will pass lightly over almost all of that history. If you'd like to learn more about the early history of nuclear physics and its unanticipated, world-changing spin-offs, I highly recommend the book *The Making of the Atomic Bomb* by Richard Rhodes.

The central discovery in nuclear physics, prior to quantum chromodynamics, was that it is useful to model atomic nuclei starting with protons and neutrons as ingredients. But some new force had to act among those ingredients, to hold the nucleus together, since electrical repulsion among the protons wants to blow it apart, and gravity is far too weak. People called this new force the strong force, and they set out to understand it. When people investigated the behavior of protons and neutrons with that goal in mind, however, things got very messy very quickly. Decisive progress occurred only after they looked *inside* protons.

Inside Protons

To look inside protons, physicists follow a similar strategy to the ones they used earlier to study the interiors of atoms—scattering experiments, à la Geiger and Marsden, which we discussed earlier, but with different kinds of beams and with an added refinement. They expose the subject of our attention to a beam of particles, watch to see how those particles get deflected, and from that observed pattern of effects work backward to the structure that causes them.

The crucial refinement is that one must study not only how much the beam particles (which in the pioneering experiments were electrons) get deflected, but also how much energy they lose. That extra information allows us to get resolution in time as well as in space. It allows us, after a lot of image processing, to get *snapshots* of proton interiors. It's important to get snapshots, it turns out, because inside protons things are moving fast. Long exposures—which in this context mean exposures longer than a millionth of a billionth of a billionth of a second—show only a blur.

Freedom and Confinement

Pictures of proton interiors revealed several surprises. They showed, first of all, that protons contain smaller particles, including quarks. Quarks had previously been used by scientists as a theoretical tool for organizing observations about strongly interacting particles, but their physical existence was widely doubted. Even one of their inventors, Murray Gell-Mann, expressed doubts. He compared his quarks to the veal in a French recipe, where "a piece of pheasant meat is cooked between two slices of veal, which are then discarded."

(The other inventor of quarks, George Zweig, took them much more literally. He spent many years trying to devise ways to detect isolated quarks, outside of protons. Those attempts never panned out, and now we know—or think we know—that they were doomed to failure.)

Skepticism about the existence of quarks was not unreasonable before their observation, because they have some unprecedented properties and behaviors. For one thing, their electric charge is a fraction of an electron's. Fractional charges had never been encountered before. For another, quarks are never found in isolation, but only within protons and other strongly interacting particles (so-called hadrons).

The latter behavior, called "confinement," continued to be puzzling, even after the quark-revealing snapshots of protons came to light. Inside the proton, it appeared, quarks hardly affected one another's behavior. Yet ultimately the forces between them must prevent any from escaping.

My first mature research in physics, done as a graduate student with my adviser, David Gross, addressed that problem. We wanted to find a theory that explained that paradoxical behavior of quarks but retained the "sacred principles" of locality, relativity, and quantum theory.

Thus, we hunted for a theory *based on quantum fields* that leads to forces between particles that are powerfully attractive when the particles are far apart but grow feeble as the particles come together. In everyday life we can manufacture such forces from rubber bands. But rubber bands are not quantum fields. Getting quantum fields to act like rubber bands is not so easy.

After a brief but intense struggle, we found a theory that does the job. It is the theory called quantum chromodynamics, or QCD. At first, the evidence for our theory was very tenuous.

But over time, as people performed experiments at higher energies and used computers to solve more problems, the evidence began to accumulate and solidify. By now, almost fifty years later, it is mountainous.

It has been a transcendent gift to experience each step on a path leading from vague aspirations and puzzlement through disciplined exploration, glimmers of enlightenment, calculations, testable predictions, and finally, at journey's end, to shared truths about physical reality. David Gross and I received the Nobel Prize for our work in 2004. We shared it with David Politzer, who did related calculations independently.

Mass from Energy: $m = E/c^2$

Now I'll discuss one of QCD's most striking applications. QCD explains the origin of most of our mass.

Einstein's famous formula $E = mc^2$ expresses the energy latent in an object at rest, due to its mass. Since energy is conserved, we can use that formula to calculate how much energy is liberated when a particle breaks up or decays into particles of smaller mass. This formula gets used in that way when we trace how energy from Earth's radioactivity moves continents (plate tectonics), for example, or how nuclear burning powers stars.

It is a beautiful thing that the logic of the formula can also be read in the opposite direction, to produce mass from pure energy: $m = E/c^2$. This is, in fact, how most of the mass of protons and neutrons—and thus the mass of human beings and the objects of everyday life—emerges.

Inside protons we have quarks and gluons.* Quarks have very small masses, and gluons have zero mass. But inside protons they are moving around very fast, and thus they carry energy. All that energy adds up. When the accumulated energy is packaged into an object that is at rest overall, such as the proton as a whole, then that object has the mass $m = E/c^2$. This accounts for almost all of the mass of protons and neutrons, as a product of pure energy. Almost all of the mass of human beings, in turn, arises from the mass of the protons and neutrons they contain. Mystics, especially in the Chinese tradition, often speak of *chi*, a universal energy that flows through creation, and they try to cultivate their inner *chi*. QCD teaches us that we come by it naturally.

One of my earliest childhood memories is a of small notebook I kept when I was first learning about relativity, on the one hand, and algebra, on the other. I didn't really understand either subject, but I thought that if I worked at it, I might discover something wonderful, like $E = mc^2$. I had $m = E/c^2$ in that notebook. Little did I know . . .

GRAVITY (GENERAL RELATIVITY)

Newton's Coincidence

Newton's theory of gravity, based on the simple force law we described previously, went from success to success for more

* And also a small proportion of antiquarks—that's a complication I'll spare you here.

than two hundred years. From the beginning, though, it contained a striking, unexplained coincidence—actually, an infinite number of coincidences. According to Newton's laws of motion, the force exerted on a body equals the body's mass times the acceleration that the force induces. On the other hand, according to Newton's law of gravity, the force exerted on a body is *also* proportional to that body's mass. Putting those two laws together, we see that the body's mass cancels. Gravity, in other words, provides a universal source of acceleration, the same for every object it acts upon.

There are two distinctive kinds of mass in Newton's theory. In one context, inertial mass governs a body's response to forces in general. In another context, gravitational mass governs the gravitational force that a body feels or exerts.* There is nothing in the theory's logical structure which requires that inertial mass and gravitational mass are proportional. The theory would still function perfectly well if that were not the case. One could imagine, for instance, that the ratio of inertial to gravitational mass might depend on a body's chemical composition. Newton's theory left the never-failing proportionality of inertial and gravitational mass or, equivalently, the universality of gravitational acceleration, as an unexplained coincidence.

Responsive Space-Time

Einstein put forward his theory of gravity, the general theory of relativity, in 1915. It explains Newton's coincidence in an

---

* According to Newton's third law of motion, action = reaction, the felt force is equal in magnitude to the exerted force.

astonishing and deeply satisfying way. It also fulfills Newton's aspiration for a theory of gravity based on local action, by bringing gravity within the same field-based framework as electromagnetism.

If we don't insist on mathematical details—and here, of course, we won't—then we can portray the majestic logic of general relativity in ten broad strokes:

1. A universal truth should have a universal explanation.
2. Therefore, the "coincidence" that gravity will impart the same acceleration to any body that occupies a given position at a given time, regardless of the body's properties, should be foundational.
3. Thus, gravitational acceleration should reflect a property of space-time.
4. One property that space-time can have is curvature.*
5. The curvature of space-time affects the motion of bodies moving in space-time. Bodies that move "as straight as possible" might nevertheless fail to move in a straight line.
6. In space-time, a straight line represents motion at a constant velocity. Deviation from straight-line motion, therefore, represents acceleration.
7. Combining points 5 and 6, we see a way to achieve point 3: Gravity reflects space-time curvature.

* Here we consider space-time as a geometric object. Adding time to space results in a geometric object—space-time—that has one more dimension than space itself, but can still be discussed using geometric concepts.

8. Since curvature can vary from place to place, and in time, it defines a field.
9. To have a theory of gravity, we need to have an equation that connects the curvature field of space-time to the influence of matter. Indeed, as Newton taught us, matter can exert gravity.
10. Newton's law of gravity suggests that the crucial property of matter, in exerting gravity, is its mass. It suggests, more specifically, that space-time curvature, which encodes gravity, should be proportional to mass. That suggestion is on the right track. It must be refined in order to get a precise equation, but the necessary refinement, once you have special relativity, is a matter of technique. (As I mentioned earlier, the main refinement is to recognize that all forms of energy, and not only mass-energy, exert gravity.)

John Wheeler, the poet of relativity, summed it up this way: "Space-time tells matter how to move; matter tells space-time how to bend."

THE WEAK FORCE

Natural Alchemy

The weak force neither binds things together nor moves things around. Its importance lies in its power to transform. Its transformative power, *leveraged by its very weakness*, gives it a unique, central role in the evolution of the universe. The weak force supplies a kind of cosmic storage battery, allowing for the slow release of cosmic energy.

In getting acquainted with the weak force, the process of neutron decay is a good place to start. It is one of the simplest weak force processes, and also one of the most important. Isolated neutrons decay with a half-life of a little over ten minutes, almost always into a proton, an electron, and an antineutrino. (Antineutrinos are the antiparticles of neutrinos.) Since neutrons and protons are much heavier than the other particles, another perspective on neutron decay can be illuminating. We can think of it as the conversion of neutrons into protons, with release of energy.

The first thing to notice is that ten minutes, in the subatomic world, is an eternity.

By way of comparison, the lifetimes of hadrons that decay through strong interactions, by reshuffling quarks and gluons, are tiny fractions of a second. The strong force acts about 10^{27}, or 1,000,000,000,000,000,000,000,000,000, times faster. By that standard, the instability introduced by the weak force, which causes neutron decay, takes a very long time to build up and become effective. In other words, it is a very weak instability. That is why we refer to its cause as the *weak* force.

The elementary particle process that underlies neutron decay is the transformation of a *d* quark into a *u* quark (plus an electron and an antineutrino). Since neutrons are based on the quark combination (*udd*), while protons are based on the quark combination (*uud*), that transformation of quarks serves to transform neutrons into protons.

Although the weak force is feeble, it can do things that the other forces can't. Neither the strong force, nor the electromagnetic force, nor gravity can change one kind of quark into another kind. The weak force, on the other hand, has the ability to transform heavier quarks into lighter ones. All the "bonus particles" we mentioned in the previous chapter* are highly unstable, due to the weak force.

The weak force acts upon quarks wherever they are. And so, specifically, the weak force can transform neutrons into protons not only when the neutrons are isolated, but also when they are within an atomic nucleus. After that happens, the new nucleus has one more proton and one less neutron than the old one. (The electron and the antineutrino escape.) Since the number of protons in an atomic nucleus ultimately determines the electrical character of the atom, and thus its chemistry, our process changes an atom of one chemical element into an atom of another. That is the sort of thing that alchemists aspired to do, but which the pioneers of modern chemistry said could not be done. The weak force performs natural alchemy.

THE FUTURE OF COMPREHENSION

Is That All There Is?

Already in 1929, Paul Dirac, the great mathematical physicist who removed the guesswork from quantum electrodynamics,

* And discuss in the appendix.

declared, "The underlying physical laws necessary for the mathematical theory of a large part of physics and the whole of chemistry are thus completely known."

Dirac was referring to the laws of quantum electrodynamics, applied to matter assumed to be made from electrons, photons, and atomic nuclei. Through ninety years hosting thousands of new experiments, applications, and discoveries in atomic physics and chemistry, Dirac's bold claim has not only survived, but become even truer, as the theory became more rigorous. And as the strong and weak forces came to be understood, the scope of fundamental understanding expanded—"a large part of physics" got much larger. The physics of 1929, for instance, had no clear ideas about how stars derive their energy or about what forces hold atomic nuclei together. Today, we know those things with confidence, thanks to thousands of stringent experimental tests.

When Dirac continued, "And the difficulty lies only in the fact that application of these laws leads to equations that are too complex to be solved," modern supercomputers were not even a dream. With their help, we're getting much better at solving the equations that fundamental understanding has provided for us. The equations of QED, QCD, general relativity, and the weak force, working in the framework of quantum theory, have powered many advances, including lasers, transistors, nuclear reactors, magnetic resonance imaging (MRI), and GPS.

Chemists and materials engineers won't be going out of business anytime soon, though. Once we go beyond a few simple cases, involving small molecules or perfect crystals, it isn't practical to predict behavior though brute force calculation. Chemists and engineers rarely if ever deal with quarks and gluons. To make progress, people must invent approximations; introduce idealizations; build faster, more powerful computers; and do experiments.

It's a different question, though, whether "the difficulty lies *only* in the fact" that our fundamental equations can be hard to solve. Might there be big effects that they are missing altogether—or is that all there is?

Our laws for the four fundamental forces, taken together, comprise what is sometimes called the "Standard Model" or (my preference) "the Core." They work together like a well-oiled machine. There are good reasons to think that the Core—our fundamental laws for QED, QCD, gravitation, and the weak force, taken together—forms an adequate foundation for practical applications of physics and that it will remain the foundation for the foreseeable future.

One reason is straightforward. The laws have now been tested with far greater precision and in a far wider range of conditions than are needed for practical applications in chemistry, biology, engineering, or even astrophysics (apart from early universe cosmology).

Another reason is more theoretical. Quantum fields are powerful tools, but they are ornery ones. It is devilishly hard to use them in a mathematically consistent way. If you're not careful, you will stumble into systems of equations that have no solutions. This gives the Core, which is heavily invested in quantum fields, a kind of rigidity. It is difficult to modify the Core without utterly wrecking it.

You can *add* to the Core, but the additions must either involve new forms of matter that couple feebly to the matter we know, or else only modify behavior of elementary particles at "impractical"—that is, very high—energies. Axions, which we'll discuss later, are an example of the former. Superstring theory, which postulates that our elementary particles are actually strings, is an example of the latter.* These sorts of additions might help to relieve the cosmological and aesthetic shortcomings of our fundamental equations, but they are unlikely to affect any of their practical applications.

To paraphrase Dirac: That's all there is, *for practical purposes.*

Thankfully, though, there's more to life than laying foundations—or being practical.

Unifying the Forces

The Core contains the seeds of its own transcendence.

Three of the four forces—QED, QCD, and the weak force—are based on different kinds of charges.† We have fields that respond to the charges, and fields that can change some of the charges into others. (Color gluon fields change one kind of color charge into another, for example.) We have electric charge, three kinds of color charge, and two weak charges. What could be more natural than to imagine a larger framework, which treats all of those charges on the same footing, and allows transformations among all of them?

* The hypothetical strings are both very tiny and very stiff, so they're both hard to discern and hard to excite.

† This aspect of the weak force is discussed in chapter 8.

That attractive idea faces a big problem: There is absolutely no evidence that the desired transformations are possible. On the contrary, they must occur very rarely, if at all. If it is possible to transform color charges into the other forms, then quarks will be able to change into electrons, and protons will be unstable. But people have looked very hard for proton decay, and they have never observed it.

On the other hand, we have learned, in the theory of the weak interaction, a way to salvage beautiful equations that seem "too good for this world." We can imagine an emptier world where the more beautiful equations hold, and then make it our world by filling it with an appropriate substance (the Higgs condensate).*

Can we take that strategy further? Might the differences among the charges be due to the complicating influence of other cosmic media, made from heavier and more elusive Higgs-like particles?

There is a beautiful reason to think so. It arises out of another key idea from the Core: asymptotic freedom. Asymptotic freedom is the weakening of the strong force at short distances. We discussed it earlier, without naming it. Asymptotic freedom was the key to discovering QCD, and it is the source of much of QCD's predictive power. We can also calculate, using the same techniques, how the other forces change with distance. Those calculations lead to a marvelous result. We find that at extremely short distances, unification is achieved. The strengths of all four forces become equal. This is exactly what we predict

---

* We'll explore this more fully in chapter 8.

to happen, in the unified field theory. By looking at short distances we minimize the effect of the complicating medium. There we seem to glimpse, in calculated numbers, the ideal world we imagined.* In this way, Einstein's vague dreams of a unified field theory have become specific, and even quantitative.

The vision that fuels our drive toward unification is a natural, logical extension of central ideas from the Core: equations based on charges and their transformations, symmetry obscured by world-filling media, asymptotic freedom. Working together, these ideas explain a "coincidence" among the strengths of the forces (including gravity). If and when people observe proton decay, this vision will be vindicated. The search continues.

Seeing Things Whole

> The objective world simply *is*, it does not *happen*. Only to the gaze of my consciousness, crawling along the lifeline of my body, does a section of this world come to life as a fleeting image in space which continuously changes in time.
>
> —*Hermann Weyl*

The idea that "the basic laws describe change" served, in our previous chapter, as the first guiding principle that leads to scientific understanding of how the world works. It has served

* Full disclosure: These calculations involve extrapolating the laws well beyond where they have been tested, and the agreement is only approximate. A more conservative way to state the situation is that the calculations work well enough to establish a suspicious "coincidence."

us well. The fundamental laws of the Core have that character. They tell us what *happens*.

But the boundary between what *is* and what *happens* is not entirely hard and fast. Eternal laws of change do not themselves change. They do not come to be, but simply are. And by drawing out their consequences, we can say a lot about enduring features of the world—or, in other words, what *is*—even though, on the face of it, they say only what *happens*.

For example, when you ask what *happens* when you examine matter minutely, and discover that matter *is* made from a few ingredients, each with a few simple properties, you've crossed that boundary. When you ask what *happens* when you bring those ingredients together and let them settle down, and discover that matter *is* organized into the nuclei, atoms, and molecules that fill out the periodic table and the reference manuals of physics and chemistry, you've crossed it again.

Still, the laws of the Core must be informed about the state of the universe at *some* time, before they can get about the business of constructing a world. They do not capture the God's-eye view, which sees space-time as a whole, all at once. Their working material is not what Weyl called "the objective world," but only slices of that world.

General relativity teaches us that the separation of space-time into space and time is unnatural. Big bang cosmology, which we'll take up in chapter 6, teaches us that the universe was remarkably simple, early on. These are big hints that we should look for more encompassing laws that will see things whole.

5

THERE'S PLENTY OF MATTER AND ENERGY

In earlier chapters, we explored the abundance of space and of time. We reached, in both cases, four fundamental understandings. First, that the universe contains overwhelming riches. Second, that in practice only a tiny fraction of those riches is available to us. Third, that the fraction we are given remains, for human purposes, plentiful. And fourth, that we are far from fully exploiting what we are given. There is still plenty of room for growth.

In this chapter, we will explore the abundance of matter and energy. Here, too, we will arrive at those four fundamental understandings.

THE ABUNDANCE OF COSMIC ENERGY

Let us begin with some comparisons, to get the measure of cosmic energy on human scales. A typical human adult takes in about 2,000 calories daily. That is roughly enough energy to run a 100-watt light bulb continuously. Over a year, it amounts to 3 billion joules. (A joule of energy, by definition, supplies a watt of power for one second, and there are about 30 million seconds in a year.) Let's call that quantity of energy an AHUMEN—pronounced, of course, "a human"—for Annual Human Energy. Of that amount, about 20 percent is used to support brain activity.

In 2020, world energy consumption was approximately 1.9×10^{11}—that is, 190 billion—AHUMENs. Since the world population in 2020 was about 7.5 billion, that amounts to roughly 25 AHUMENs of energy consumed per human. This number, 25, is the ratio of total energy consumed to the amount of energy used in natural metabolism. It is an objective measure of how far humans have progressed, economically, beyond scratching out a bare subsistence. Americans, for comparison, consume roughly 95 AHUMENs per person.

The annual energy output of our Sun is enough to supply roughly 500 trillion AHUMENs to each human. You should not fail to notice that 500 trillion is a great deal larger than 25, or even 95. Thus, fundamentally, there's vast room for economic growth based on harvesting a larger fraction of our Sun's energy output.

Of course, the Sun's output gets radiated in all directions. To capture a bigger fraction, we'd need to make significant investments of time and resources to put gigantic collection devices in space. Freeman Dyson and others have proposed engineering projects of that sort, called Dyson spheres.

If, more modestly, we restrict ourselves to the portion of

solar energy that makes it to Earth, then we find "only" about 10,000 times our present total energy consumption. That number provides a more realistic baseline from which to assess the economic potential of solar energy. Evidently, even without a Dyson sphere, there's still *plenty* of room for growth.

Here we have considered the energy emitted from our Sun. Earlier, in our survey of the universe, we came to see our Sun as just one star among many. With that in mind, we understand that the universe as a whole is awash in vastly more energy than humans will, for the foreseeable future, be able to access. What we can do, though, is capture tiny samples from those dispersed riches. That's what astronomy is all about. Astronomy enriches our minds, if not our economy.

This discussion of comparisons gives objective meaning to the claim that there's plenty of matter and energy. There's more than enough to make objects as complex and dynamic as humans and to support an extremely expansive human agenda.*

FUNDAMENTALS AND HUMAN PURPOSES

Dynamic Complexity

By making simple comparisons, we have demonstrated that there is, for human purposes, plenty of energy in the universe.

---

* I have *not* described how humans actually arose, historically, nor have I described what the human agenda is, or should be. Those are grand subjects, but they belong in books different from this one.

Now let us consider, from a more fundamental perspective, *why* there is.

To do that, we must address two basic questions:

> What is it, in the physical universe, that embodies "human purposes"?
>
> Why does realizing that thing require so little energy, compared to what our Sun puts forth?

The first question can be addressed at many different levels. If we try to define "human purposes" precisely, we risk a rapid plunge into murky depths of vague metaphysics. But if we ask what it is that is essential to what people do, and to what they are, in physical terms, then the answer that emerges is clearer than the question is. At that level, the heart of the matter is *dynamic complexity*. Although there's no scientific consensus on precisely how to define complexity, we "know it when we see it," in examples like these:

- To learn and to think, we change patterns of connections, secretions, and electrical impulses in our brains. To sense the world, we transform incoming patterns of electromagnetic radiation (sight), air pressure (hearing), local chemistry (taste and smell), and a few other data streams into that common brain currency. To move and to act on the world, we use muscle power, ultimately based on the synchronized contraction of well-organized protein molecules.
- In building temples, synagogues, mosques, or cathedrals,

people draw up plans, gather materials, use construction tools and machinery, and employ builders and artists to create complex, "unnatural," "spiritual" environments where none existed before.

- Music and ritual are purified expressions of dynamic complexity.

Each of those quintessentially human activities involves, at its core, complex material patterns that change in time. In different cases, the patterned matter takes different forms, ranging from neural networks to vibrations in air; and it embodies different things, including tools, symbols, memories, signals, instructions, and actors. Dynamic complexity is the deep structure underlying them all.

Here on Earth, through most of biological and human history, the physical realization of dynamic complexity has hinged upon making and breaking enormous numbers of chemical bonds, using power supplied by the Sun. Today, other possibilities are opening up, as I'll discuss below. But Sun-powered making and breaking of chemical bonds is still the central method, and we should discuss it first.

EXPLOSION BY CONSTRUCTION

Atoms have many features that make them excellent pieces with which to build up interesting and intricate—that is, complex—creations:

- There are many kinds of atoms, one for each chemical element. All the atoms of any particular element are essentially identical.* Thus, they provide a wide stock of interchangeable parts.
- Atoms are available in enormous numbers. A typical human body contains an octillion or so, which is more than the number of stars in the visible universe.
- Atoms can combine together into bigger units—molecules—following the rules of quantum theory and the laws of electrodynamics. We say that the atoms are joined by chemical bonds to make a molecule.

To understand how those fundamental facts can lead, under favorable conditions, to dynamic complexity on a grand scale, we need to bring in two big ideas: combinatorial explosion and provisional stability.

Combinatorial explosion, in its simplest form, is the rapid growth in the number of overall possibilities as you make several independent choices. Thus, if I can choose any one of ten digits to fill nine different places, then I can make 10^9, or one billion, different combinations—namely, the numbers 000000000, 000000001, 000000002 . . . 999999999. Ten and nine are reasonably small numbers, but 10^9 is quite a large one. This demonstrates the essence of combinatorial explosion.

In DNA, we get to make four choices among nucleotides

---

* Some elements can occur as a few different isotopes. Here the atoms have similar chemical properties, but they have different numbers of neutrons in their nuclei. We encountered an example earlier, in chapter 2, when we discussed carbon dating.

(guanine, adenine, thymine, cytosine—G, A, T, C) to attach at each spot along a long sugar–phosphate backbone, and there can be many thousands of spots. Proteins, similarly, involve choosing among twenty amino acids attached to stereotyped backbones of variable length. Those architectures support combinatorial explosions of precisely the same type as the decimal expansion of numbers, but in base 4 or base 20. Thus, DNA sequences, which are used to store information, can record enormous quantities of information. And proteins, which provide the structural and functional building blocks for life, form a huge inventory. Different proteins fold into an enormous variety of sizes and shapes, with diverse mechanical and electrical properties.

Molecules of other kinds, in both the organic and the inorganic worlds, can branch, form loops, agglomerate into membranes, stack regularly into crystals, and do many other tricks. This wealth of possibilities leads to a combinatorial explosion of combinatorial explosions. When you fold in the fact that a single gram of matter contains billions of billions of atoms, it becomes clear that there's no shortage of material to support complexity on a grand scale. William Blake's poetic description of an "infinity in the palm of your hand" has a sound scientific basis.

CONCEIVING COMPLEXITY

To deliver on that material's potential, we must be able to sculpt it. We want our atomic building blocks, like Lego bricks, Tinkertoys, or the ball-and-stick models of atoms and molecules used in chemistry classes, to click together easily, to come apart easily, and to stay put in between. This key property, *provisional stability*, requires a nice balance between stability and changeability.

Chemists work to determine *what's realistically possible* in the world of molecular complexity, and biologists work to determine *what actually happened.* The work of chemists and biologists is open-ended and endlessly fascinating. I will rely on their goodwill and sense of humor to indulge my drastic simplifications. What can be understood reasonably simply, and what I will describe here, is only how the world, and specifically the Sun-Earth system, "conspires" to make intricate sculpting of matter *conceivable.*

Three crucial ingredients make provisional stability possible. They are a high temperature, a low temperature, and an intermediate energy scale. The high temperature is the temperature at the surface of the Sun, around 6,000°C. The low temperature is the temperature at the surface of Earth, around 20°C. The intermediate energy scale is the quantity of energy it takes to make or break a typical chemical bond, which is roughly an electron volt.

Temperatures around 20°C leave molecules mechanically flexible, but they don't often break chemical bonds, because the energies they supply rarely reach an electron volt. On the other hand, photons arriving from the surface of the Sun pack more concentrated energy, often exceeding an electron volt. They are capable of breaking chemical bonds. The interplay between that cool, but not frigid, background and that accessible, but not oppressive, supply of concentrated energy makes it possible, but not too easy, to rearrange molecular patterns. This sort of provisional stability available on Earth is just what we need, physically, for dynamic complexity.

To complete our story of abundant potential for dynamic complexity and how it gets realized on Earth, we need to understand, based on fundamentals, how our Sun manages to fulfill its role. But before turning to that, let us pause to calibrate our own dynamic complexity.

The basic units of human brains are neurons. The number of neurons in a human brain is roughly one hundred billion, or 100,000,000,000, or 10^{11}. While well below an octillion, this is still an unimaginably large number. It is roughly equal to the number of stars in our galaxy.

Each neuron is an impressive little information-processing device. Individual neurons are wired together through many connections. Typical neurons can make hundreds or even a few thousand connections to other neurons. Much of what we learn is encoded in the varying strength of these connections, as useful patterns of influence get reinforced and useless ones whittled away. Peak connectivity occurs between the ages of two and three, but peak complexity occurs later, after a lot of selective whittling.

If we consider the possible ways for that many neurons with that many connections to get wired up, we get into dizzying numbers, well beyond octillions. Our skulls host mind-blowing combinatorial explosions. We should not be shocked to find that this unimaginably large number of neurons, wired in such unimaginably intricate patterns, working together, can do astonishing things. Walt Whitman really did contain multitudes. So do I. So do you.

FUEL TO BURN, *SLOWLY*

The Sun runs on nuclear fuel. It is a giant fusion reactor. The nuclear burning process that drives the Sun is the conversion of hydrogen into helium. A hydrogen atom contains one proton and one electron. A helium atom contains two protons, two neutrons, and two electrons. In the Sun, a chain of reactions results in the conversion of four hydrogen atoms into one helium atom plus two neutrinos, releasing energy.

If you recall our discussion of neutron decay in the preceding chapter, you might think there's been a typo just now. There we saw that isolated neutrons want to turn into protons. That decay process liberates energy, because neutrons are slightly heavier than protons. In our description of solar burning, we've got the opposite happening—protons turning into neutrons. But it's not a typo. In a helium nucleus there are powerful attractions among the protons and neutrons, due to the strong force. By bringing together the separate pieces, one gains a lot of energy. Thus, protons can turn into *bound* neutrons, with energy to spare.

Transformations between protons and neutrons, in either direction, require the weak force. That makes neutron decay a slow process, by particle physics standards, as we discussed earlier. In the Sun's nuclear burning, the slowness of the weak force gets greatly amplified. In that burning process, one must bring the particles together before transforming them. But those close encounters are fleeting, so "quality time" accumulates very slowly. It takes billions of years, on average, for protons in the Sun to convert into (bound) neutrons. Thus, thankfully, the Sun's fuel supply will last for several billion more years. On the other hand, the amount of hydrogen in the Sun is so enormous that even this slow burning is enough to keep it shining.

SUMMING UP: THAT ART THOU

This completes our account of how dynamic complexity arises on Earth, from the perspective of physical fundamentals. It grounds biology, and ultimately psychology and economics, within our deep understanding of material reality.

Each of the four fundamental forces plays a different, crucial role in this story. Gravity keeps Earth in orbit around the Sun, at a nice distance, where the equilibrium temperature supports dynamic complexity. The electromagnetic force, QED, weaves atoms into molecules. The strong force, QCD, supplies the attractions that make nuclear burning possible. The weak force enables the transformations that allow nuclear burning to proceed, but only slowly.

THE FUTURE OF MATERIAL ABUNDANCE

New Places, New Pieces, New Minds

The principle that the essence of human purposes is expressed through flows of information in dynamic complexity, rather than through details of chemistry and physiology, is both mind-expanding and liberating. It challenges us to imagine how minds could emerge elsewhere in the universe, and it prepares us to embrace those minds within our circle of empathy.

To thrive, human bodies require specific conditions, including temperatures within a narrow range, air that contains a special mix of molecules and is free of toxins, a reliable supply of water and nutrients, and protection from ultraviolet radiation and cosmic rays. These conditions exist within a thin layer near the surface of Earth, but they are very rare within the universe as a whole. Colonization of space by humans, in our Earth-adapted bodies, is a crazily difficult project.

Expanding the sphere of influence of human information is a much easier, more realistic goal, and it is no less meaningful. The actuators and sensors we send can create and explore on our behalf—and stay in touch.

Our profound understanding of matter gives us several ways to manufacture large-scale dynamic complexity that are quite different from making and breaking chemical bonds. We can supplement, or even replace, chemistry with electronics and photonics.

Digital photography is a convincing, mature example of how that occurs. Here the primary sensors—charge-coupled devices, or CCDs—count electrons liberated by photons and record the resulting numbers in arrays of 0s and 1s, encoded using any of the formats described earlier. This information, which encodes the image, can be processed in many ways; for example, to remove noise, highlight interesting features, or otherwise beautify the picture. Then, after processing, you can translate the information back into images, by using it to instruct displays. All that processing is done electronically, in computers or specialized chips. Photographic plates, emulsions, and darkrooms, which once gave photography an aura of romance and mystery—while making it much more time-consuming and difficult—are on the wane.

The evolving patterns of connection and chemistry-driven electrical activity in human brains are the apex of dynamic complexity, and of mind, today. But the importance of other embodiments of dynamic complexity is increasing, and there's plenty of room for it to grow.

Inside modern computers, information is stored and processed in arrangements and rearrangements of electrons, as opposed to entire atoms or molecules. The energies involved can be much smaller, and the processing can be much faster. To represent information, we have either a high concentration of electrons (leading to a low voltage, interpreted as "0"), or a low concentration (leading to a high voltage, interpreted as "1") in each of billions or trillions of tiny buckets. In this way, we manufacture a combinatorial explosion of provisionally stable units. It is a versatile platform for dynamic complexity.

It is also possible to use the direction of electron spins—up or down—instead of the electrons' concentration, to embody 0 and 1. Manipulating spin directions is more delicate work than pushing charge around, but in principle it can be faster and more energy efficient. We can also work with photons instead of electrons, and monitor their concentrations (amplitude), colors (wavelength), or spin (polarization).

These post-chemical platforms for dynamic complexity have big advantages in speed, size, and energy efficiency. They are also more open to controlled exploitation of the richness of the quantum world.* They can support continued growth of mind in the cosmos for a long time, and on a vast scale.

How Things Could Go Wrong

> With great power comes great responsibility.
>
> —*Peter Parker (Spider-Man)*

An overarching message from our fundamentals is that there's plenty of space, plenty of time, and plenty of matter and energy. The physical world offers us humans a future much bigger, longer, and richer than what we've achieved so far—if we don't blow it.

Many things could go wrong. Plagues have ravaged human civilizations in the past and caused significant setbacks, as have earthquakes and volcanic eruptions. An unfortunate collision

* The full quantum-mechanical description of a system is much more elaborate than its classical description. We'll explore this more deeply in the final chapter. This gives us, in principle, a bigger sketchpad—but one that is strange and hard to work with. Quantum information technology is a research frontier.

of Earth with cosmic debris doomed the dinosaurs. We can and should work to mitigate those dangers. But here, to close this chapter, I will briefly highlight two possible human-made failure modes that loom large today, and that are closely connected to its themes.

Our Sun supplies Earth, at a steady rate, with far more energy than humans presently use. Technology to capture a larger fraction of that energy is developing rapidly, and there is little doubt that in the foreseeable future—barring catastrophe—we will be able to use it to support a richer world economy, sustainably.

At the moment, however, it is easier and more convenient to tap into solar energy that was captured long ago by plant life, and now is stored in fossil fuels—coal and oil. Unfortunately, burning those fuels on a large scale releases enough carbon dioxide and other pollutants into our atmosphere to alter its properties. The polluted atmosphere traps more of the Sun's energy, causing Earth's average temperature to rise. This is the first human-generated crisis looming over us.

Our sister planet, Venus, is a jewel of the night sky. It is a warning beacon, too. Its atmosphere, rich in carbon dioxide, traps the Sun's energy extremely efficiently. Surface temperatures on Venus approach 460°C (860°F), which is hot enough to melt lead, and precludes complex chemistry. Venus is closer to the Sun than is Earth, but if we put it at Earth's orbit, its temperature would still be alarmingly high—about 340°C (645°F). Earth won't get that hot anytime soon, but even a few degrees of added temperature will have drastic, possibly catastrophic effects. Rising temperatures are causing polar ice to melt, leading to rising sea levels; violent weather patterns are emerging, driven by increased atmospheric moisture; and we are disrupting the lives of temperature-sensitive plants and animals, thus endangering our food supplies (and our friends).

The second human-generated threat is nuclear weaponry. As scientists explored the strong and weak forces, they discovered potent new fuels based on nuclear rather than chemical burning. Famously, this enabled the construction of new sorts of bombs, with much greater destructive power. Were a significant fraction of those bombs to be used in warfare, many millions of people would perish in horrible ways, and important centers of civilization would become uninhabitable wastelands. Human progress would be set back catastrophically, and perhaps irreversibly.

The blessings of economic growth and scientific knowledge come together with severe dangers. Those dangers *can* be avoided. Whether they *will* be is an open question.

II

Beginnings and Ends

6

COSMIC HISTORY IS AN OPEN BOOK

Our first five fundamentals have described the basic ingredients of physical reality: space, time, fields, laws, and dynamic complexity. They addressed "what there is." Our next two will address "how it got this way."

People have speculated about the origin of the physical world ever since there have been people. Anthropologists have recorded creation stories from many cultures. Literature contains many others, some of which have, at different times and places, been accorded sacred authority. But adequate intellectual and technical tools to address the question of physical origins first became available in the twentieth century.

Over the past few decades, a remarkably clear picture of the broad outline of cosmic history has emerged. The crucial breakthrough was Hubble's work on the distance and motion of galaxies. Hubble discovered that distant galaxies are moving away from us, with velocities proportional to their distances.

That universal expansion, run backward in time, suggests that the matter in the universe was once much more densely packed together, and that the universe once looked quite different from what we see around us today.

What was it like? In the text of this chapter I will address that question, in three steps. First, I'll present a bold guess about the early state of the universe, commonly known as the big bang theory. I will emphasize its *strange* simplicity. Second, I'll sketch the cosmic history which follows from that guess. Finally, I'll discuss some of the main observable consequences that flow from this history, and the evidence that has accumulated for it. The multifaceted success of this hypothetical history justifies the bold guess that launched it.

That said, the observational evidence thins out, and our equations cease to be reliable guides, when we look toward the very beginning. At the close of this chapter, I will discuss promising prospects, both theoretical and observational, for seeing deeper.

SCOPE AND LIMITS

> The work will teach you how to do it.
>
> —*Anonymous (quoted in a fortune cookie)*

Science often resembles the game of Jeopardy!, where answers suggest what the right questions are. The great mathematical astronomer Johannes Kepler, a hero in some of our earlier discussions, considered many aspects of the solar system in his work. His questions about the shape of planetary orbits and the speeds with which the planets traverse them had good* answers, now famous as Kepler's laws of planetary motion. But Kepler also wrestled with the problems of why there are six planets (as was thought at the time) and why they are at the distances from the Sun that they are. He had some amusing ideas on those subjects, which brought in music—"the music of the spheres"—and the Platonic solids. But those ideas never gelled into good answers. Today, scientists think that Kepler wasn't asking the right questions. Our fundamental laws, and our fundamental understanding of cosmic history, suggest that the size and shape of our solar system is a rather accidental feature of the universe. Its ultimate form is caught up in details of how a mess of gas, rocks, and dust collapsed and condensed to make the system we observe today. We see our solar system as one among many in the universe. In other systems we often observe different numbers of planets in different arrangements than what Kepler was hoping to explain. Since Kepler's day, too, our own solar system has grown to include Uranus, Neptune, asteroids, Pluto, and a lot of other stuff.

Cosmic history includes, in principle, an enormous range of things, including the history of life on Earth, the history of China, the history of Sweden, the history of the United States, the history of rock and roll, and so forth. But no sane person would expect to get an understanding of those subjects based on physical fundamentals.

What fundamentals-based cosmic history does provide is

* Here we say answers are "good" if they are easy to state, mathematically precise, and agree with observation.

three things. First, it offers a profoundly strange yet informative and convincing account of what the early universe was like. This account is a good answer to an interesting question, and it proves to be a rich source of surprising, observable consequences. Second, it provides a broad scenario for how the structures we observe around us—including, for example, our solar system—can have emerged. Third, it suggests exciting new questions, such as what "dark matter" is.

WHAT HAPPENED

Strangely *Simple Beginnings*

> Everything should be made as simple as possible, but no simpler.
>
> —*Albert Einstein*

As we've already discussed, Hubble's discovery, which we can loosely describe as "expansion of the universe," practically begs us to consider what happened earlier.

We seem, on the face of it, to be living out the aftermath of a universal explosion. If we can understand the beginning, then we can hope to leverage our understanding to illuminate later events.

As a first attempt to reconstruct the beginning, we can imagine "running the movie backward." To do this, in our minds, we simply reverse the velocities of all the galaxies and let the laws of physics play out.* The galaxies rush together. As they approach, they begin to attract one another gravitationally, and their accelerated motion releases energy. The matter gets mixed up and heats up. The temperature rises. Atoms get stripped of their electrons, and rapidly moving charges radiate like mad. Tightly packed, rapidly moving protons and neutrons boil into a soup of quarks and gluons. Finally, our hard-won knowledge of fundamental interaction pays off. Asymptotic freedom, in particular, implies a great simplification—at high energies, the formidable complications of the strong interaction go away. Extremely hot, dense material is surprisingly simple to understand, directly from fundamentals.

But before accepting this reconstruction of the past, we must face up to a *major* conceptual problem. The history of the universe depends on it. The problem is this: The simple picture I just sketched, as an account of cosmic expansion run backward, is desperately unstable. What we really should expect as the matter rushes together is that stars, planets, gas clouds, and whatever else is out there will merge, through the inexorable attraction of gravity, into gigantic black holes. The non-gravitational interactions do, indeed, want super-dense, energetic matter to become a hot, homogeneous gas. That is their favored equilibrium, which they will try to enforce. But gravity abhors homogeneity. Gravity wants things to clump, in general, and gravity wants super-dense matter to clump into black holes.

* Here we rely on the fact that the same fundamental laws of physics apply when we run time backward. This is very nearly—though not exactly—true. Why? That question introduces a grand mystery, which we'll take up in chapter 9.

Running our cosmic movie backward, if we didn't know better, and were honest about it, we'd "predict" that gravity wins, and the early universe devolves into big black holes rushing together and merging into bigger black holes. But if the early universe really were like that, then—running the movie forward again—we'd have a universe with essentially all the matter locked up in black holes. Once you've fallen into a big black hole, it's quite difficult to get out!

The universe we actually observe is nothing like that prediction. Our observed universe, averaged over intergalactic scales, is remarkably homogeneous. Wherever we look in the sky, if we sample a reasonably large chunk, we find the same sorts of galaxies, distributed with the same density. This was another of Hubble's pioneering discoveries. Since gravity tends to make things less homogeneous, the fact that we observe large-scale homogeneity today implies that the universe was even more homogeneous early in its history. This means, in terms of our backward-running movie, that the matter comes together "just so," in a way that is delicately orchestrated to avoid gravitational mergers.

The big bang theory of cosmic history uses the naïve picture of the early universe as a hot homogeneous gas that I sketched originally, before I raised worries about its stability. The big bang theory simply puts those worries aside. Fundamentally, therefore, the big bang theory is a strange hybrid of two opposing ideas. It postulates complete *equilibrium* for the nongravitational interactions, but maximal *disequilibrium* for gravity. Running Hubble's expanding universe backward in time suggests the former, while running Hubble's quasi-homogeneous universe backward in time suggests the latter. In the big bang theory, we follow both suggestions.

The Expanding Fireball

We start, then, with a very hot, very homogeneous gas. We also assume that space, which (according to general relativity) might be curved, is actually flat.* For a first draft of physical cosmology, *that's all we need to know.*

The ingredients in a hot gas move around so rapidly, and interact so often, that they reach a dynamic balance, known as thermal equilibrium. At the extremely high temperatures we contemplate in the earliest moments of the big bang, thermal equilibrium is especially powerful, because so many things can—and do—happen. Many kinds of particles, from photons to gluons, quarks, antiquarks, neutrinos, antineutrinos, and more, are getting produced and destroyed (or, equivalently, radiated and absorbed). In equilibrium, all are present, with predictable abundances. H. G. Wells caught the spirit of thermal equilibrium memorably: "If anything is possible, then nothing is interesting." In thermal equilibrium at extremely high temperatures, we find a completely predictable mixture of all the elementary particles.

Another aspect of super-high temperature conditions is that structures cannot hold together—molecules break up into atoms, atoms break up into electrons and nuclei, nuclei break up into quarks and gluons, and so on. In short, we get down to fundamentals.

* This is explained further in the appendix.

Given that starting point—a predictable mix of fundamental ingredients—we can use our knowledge of fundamental laws to predict what happens next. The result is simple: Our omnipresent fireball expands under its own pressure, working against its own gravity and cooling as it does.

As the fireball cools, two especially notable things happen. One is that some reactions occur more rarely, and then effectively stop. This results in lingering afterglows. For example, once the temperature gets low enough, the photons in the fireball cease to interact significantly with the other matter. In plain English, the sky clears up, so that light travels more or less freely from one end of the universe to another, as it does today. The photons that were part of the fireball don't disappear, though. They become the so-called cosmic background radiation, a lingering afterglow that fills the universe.

Another result is that particles begin to stick together. Quarks combine into protons and neutrons, electrons bind to atomic nuclei, and so forth. In this way, matter in the form we know it begins to take shape.

That is our first draft of cosmic history.

HOW WE KNOW IT

> The past is never dead. It's not even past.
>
> —*William Faulkner*

The cosmic past is never dead. It leaves relics, which we can observe today. The cosmic past is not even past. Thanks to the finite speed of light, when we receive light from far away it carries the past to us.

Reconstructing what happened in the early universe is a lot like reconstructing a crime. We survey the evidence, form a theory of the case, and look for corroborating evidence. If we find surprises, then we must refine our theory, or change it.

Cosmic Census

With better telescopes and cameras, and more powerful ways of handling data, astronomers have been able to survey the universe far deeper and more fully than was possible for Edwin Hubble. His work made the big bang a prime suspect; their work could sustain a conviction.

You may recall that Hubble discovered that distant galaxies are moving away from us, with their velocity proportional to their distance. This relationship, run backward in time, suggested the big bang. It holds accurately for nearby galaxies, but we should not expect it to work for the most distant ones. Velocity proportional to distance will not bring distant galaxies together at the same time, because (in our movie played in reverse) gravitational forces come into play and modify the motion. Given the big bang as a starting point, it is possible to predict how the expansion rate changes in time. That prediction translates into a refined projection for how the redshift of galaxies depends on their distance, which can be compared with observations. It works.*

* That is, it fits together with all the other evidence into a consistent picture.

By running the expansion backward in time, we determine what is commonly called the "age of the universe." What that phrase refers to is the length of time that has passed since the universe was a much hotter, denser, more homogeneous place than it is now—or, more loosely speaking, the time that has passed since the big bang occurred. In the earliest moments following the big bang, stars and galaxies could not have held together. But we can estimate when such structures should have begun to form. And we can also estimate the ages of some very old objects in quite different ways, using radioactivity and the theory of stellar evolution, as we discussed in chapter 2. Those very different ways of assessing cosmic antiquity are found to agree quite nicely. In short, the universe is about as old as the oldest objects within it—as it should be.

Lingering Glow

The lingering glow of photons present when the fireball first cooled enough to become transparent was first detected in 1964, by Arno Penzias and Robert Wilson. Those photons have been drastically redshifted, and now they are primarily microwave radiation (the same kind of electromagnetic radiation as is used in microwave ovens). They form the so-called cosmic microwave background, or CMB. The CMB is a snapshot of the early universe, spread over the sky in invisible "light." The big bang picture not only predicts the existence of the cosmic microwave background, but also has a lot to say about the details of its composition—specifically, the intensities of its various radiation frequencies. Here, too, the observations agree with the predictions.

Relics

As the raging fireball, including quarks, antiquarks, and gluons, cools, those particles start to stick together into protons, neutrons, and other atomic nuclei. One can calculate, within the big bang model, the relative abundance of the different nuclei that emerge. It turns out that the overwhelming majority of potential nuclear material emerges from the big bang as ordinary hydrogen (^{1}H—a lone proton) and helium (^{4}He—two protons and two neutrons). There are also small admixtures of deuterium (^{2}H—one proton and one neutron, an isotope of hydrogen), tralphium (^{3}He—two protons and one neutron, an isotope of helium), and lithium (^{7}Li—three protons and four neutrons). These different isotopes have all been detected, using the techniques of spectroscopy, to occur with the predicted abundances within appropriate "unprocessed" environments.*

All other kinds of nuclei got formed in stellar processes, at a much later stage in cosmic history. Observing and understanding their abundances is a wonderful subject, but its connection to fundamentals is less direct.

* We need to be wary here of nuclear burning in stars, whose alchemy transforms atomic nuclei, as we've discussed earlier.

THE FUTURE OF COSMIC HISTORY

Inflation

As I emphasized above, the big bang theory is profoundly strange. It assumes a starting point that is actually unstable, and postulates that matter in the early universe was extremely fine-tuned—specifically, that it was uniform—to avoid triggering its gravitational instability.

There's also another uncanny aspect, which I mentioned only in passing, because a full explanation would have interrupted my narrative.* The big bang theory assumes that space is Euclidean, or "flat." Spatial flatness is consistent with Einstein's general relativity, but not required by it. Relativity is ready to accommodate spatial curvature. We need some other idea to explain why Nature does not make use of that opportunity.

My MIT colleague Alan Guth introduced a brilliant and promising idea, which addresses those issues elegantly. He proposed that the universe underwent a tremendously rapid expansion early in its history, which he calls "inflation."

It is easy to appreciate intuitively how inflation can help with our issues. If the universe inflates, then inhomogeneities in matter are diluted, and curvature is expanded away.†

But did inflation actually happen? I'd like to think so, but

* It's spelled out in the appendix.

† A round balloon will look much flatter if you blow it up to the size of Earth.

it would be good to have more specific ideas about how it happened, and more specific evidence in its favor.

Inflation is *not* a consequence of the fundamental laws we know today. It requires something more—additional forces and fields, presumably. Andrei Linde and Paul Steinhardt proposed some forces and fields that could do it, but there is no independent evidence for them. A good model of inflation might enable us to test the basic idea more rigorously and draw out new consequences. As yet there is no such model. There's a big opportunity for discovery here.

Reaching Further Back

The cosmic microwave background is a lingering afterglow of the big bang, which gives us a direct window on the universe's early history. It arises, you may recall, from photons that were present in the cosmic fireball at the time when it first cooled down enough to become transparent. That happened about 380,000 years after the big bang. While this is impressively early, relative to the 13,800,000,000-year age of the universe, a lot of fascinating events happened earlier, and we'd like to look into those, too.

Investigating those will be challenging, but there are some real prospects for accomplishing it. For instance, there are at least two other afterglows that ought to surround us. Their origins resemble that of the cosmic microwave background. They are composed of neutrinos and of gravitons.*

* Individual gravitons are probably beyond the reach of present-day technology, but the cumulative effect of many of them, filling the sky, is a realistic target.

Since neutrinos interact feebly with other sorts of matter, and gravitons more feebly still, the fireball becomes transparent to them much earlier than it does for photons. In consequence, the lingering afterglows of neutrinos and gravitons carry messages that are much older than the ones that the cosmic microwave background carries. Gravitons, in particular, can give us a glimpse of events that occurred only small fractions of a second after the big bang. There's plenty of room for surprises there. A graviton-based snapshot could show us what was going on at temperatures and other conditions far more extreme than anything that occurs in terrestrial laboratories, or most likely anywhere else in the present-day universe. We might get to see a burst of gravitational radiation spit out by material moving rapidly, during cosmic inflation, for example.

The challenge of observing these more exotic lingering afterglows arises from the same feature that makes them so fascinating, namely that they interact very feebly with other kinds of matter. We will need new and highly sensitive kinds of antennas and telescopes in order to see them at all. Those antennas and telescopes will almost certainly bear little resemblance to the ones developed for photons. Here there is much room for creativity.

There may also be other lingering afterglows, arising from particles whose existence is not yet established. After all, the basic thing about cosmic afterglows is that they arise from particles which interact so feebly with matter that the universe becomes transparent to them.

"Dark matter" could be just such an afterglow. I, and most of my colleagues, suspect that it is. Specifically, I suspect it is an afterglow of axions. I'll be spelling this out, and making the case, in chapter 9.

The Very Beginning

Because our vision gets blinded as we approach the big bang, it is not possible to speak with confidence of "the very beginning." The concept might be misguided, or even senseless. Saint Augustine made a brilliant suggestion about this in his *Confessions*, which I suspect is on the right track. A parishioner asked Augustine, "What was God doing, before He created the universe?" Augustine records that he considered answering, "Preparing hell for people who ask too many questions." But he had too much respect for his parishioner, for himself, and for God to do that. Instead he thought hard about the problem, and he prayed for an answer, as it preyed on his mind. This led him into a deep meditation on time.

Augustine reached a conclusion about the nature of time very similar to ours, in chapter 2. Basically, he concluded that time is what clocks measure—neither more nor less. That thought led him to a better answer to his parishioner's question. Before God created the world, Augustine reasoned, there were no clocks—and therefore no time, and therefore no such thing as "before." Thus, the question "What happened before God created the universe?" when carefully considered is devoid of meaning.

The essence of Augustine's answer survives translation into the language of modern physical cosmology. Nothing precedes the origin of the universe, because in that context, time—the thing that clocks measure—has no meaning.

7

COMPLEXITY EMERGES

The physical world is complicated. Rain forests, the internet, and the collected works of William Shakespeare are all contained within it. Yet our fundamentals promise to build it all up from a few ingredients, a few laws, and a strangely simple origin.

This poses a challenging question: How does complexity emerge, fundamentally? This chapter explores that question. At its close, I'll discuss the long-term prospects of cosmic complexity and how apparent complexity can exist within profound simplicity.

HOW THE UNIVERSE GOT INTERESTING

Gravity's Tenacity

> For to him that has will more be given; and he will have abundance: and from him who has not, even what he has will be taken away.
>
> —*Mark* 4:25

> For to every one who has will more be given, and he will have abundance; but from him who has not, even what he has will be taken away.
>
> —*Matthew* 25:29

These quotations describe what has come to be called the "Matthew effect," although Mark's gospel is almost certainly earlier. Loosely stated it means "the rich get richer while the poor get poorer."

The gravitational instability that is central to the emergence of complexity in the universe is a version of the Matthew effect. Dense regions in the universe exert more powerful attractions, and thus accumulate more matter, and thus become still denser. Regions that are less dense than average, conversely, will lose to the competition, and empty out further. In this way, the density contrast sharpens over time. Small contrasts evolve into larger ones. That is the gravitational instability.

To get the most out of the big bang theory, we need to

refine our assumption that the distribution of matter early on was completely uniform. Small deviations from uniformity will do, because they get amplified by gravitational instability.

Happily, the cosmic microwave background, which gives us a picture of the universe 380,000 years after the big bang, is not quite uniform. Its intensity varies with angle, at the level of parts in ten thousand, reflecting density contrasts of similar size. The detection of such tiny nonuniformities was a triumph of experimental technique. John Mather and George Smoot shared the 2006 Nobel Prize for their pioneering work on this subject.

These tiny seeds get amplified in time, by gravitational instability. According to calculations, they are just the right size to grow, in the available time, into density contrasts large enough to evolve into galaxies, stars, and the structures we presently observe in the universe.

Why was matter in the early universe so nearly—but not quite—uniform? We don't know for sure, but there's a beautiful possibility I'd like to share with you. The theory of cosmic inflation, on the face of it, suggests a conceptual explanation for perfect uniformity, as we discussed earlier. But when we try to embody the theory within the framework of fundamental physics, using quantum fields, we discover that this isn't quite right. Quantum fields have quantum-mechanical uncertainty built into them. Because of this they can't generate perfect uniformity, though they can get close. It is possible, therefore, that a good physical implementation of inflation will convince us that the structure we observe in the universe was triggered by quantum uncertainty in the early universe.

Matter's Unfinished Business

As we discussed in our fifth fundamental, nuclear burning in the Sun is the key to dynamic complexity on Earth. The Sun is, fortunately, still evolving. It has not reached equilibrium. Yet matter, according to the big bang theory, started in thermal equilibrium. How did the material of our Sun escape from it?

We can trace the sequence of events. The cosmic fireball expanded and cooled. Thermal equilibrium requires frequent interactions, but the fireball was becoming less intense and increasingly sluggish. Eventually, thermal equilibrium started to break down.

The cosmic microwave background and other potential lingering afterglows we discussed reflect breakdown of equilibrium. Here photons—or neutrinos, gravitons, and axions—came to interact very rarely.

For the Sun and other stars, what's important is that nuclear burning during the big bang did not run to its logical conclusion. In the expanding universe, many protons could not find each other and combine, until—much later—they got brought back together, in the Sun and other stars. The *combustible* mixture of nuclei that emerges from the big bang is another of its lingering afterglows.

SENSITIVITY: THE BRANCHING OF REALITY

Dice games, bowling, and many other recreations and sports would be dull—though possibly lucrative—if you could reliably connect input to output. You could master the motions required to roll a seven, or to bowl a strike, once and for all, and be done with it. But in practice this is impossible, because small differences in muscular motions, moisture on your hand, dirt on the rolling surfaces, or many other tiny effects can change your outcome. In short, the final result depends sensitively upon many factors that are essentially impossible to predict or control.

Similarly, as gravitational instability plays out and matter clumps, the exact form this clumping ultimately takes in any particular place depends sensitively on the starting positions and velocities of many individual particles. Calculations reveal that gas clouds with only subtle differences to begin with can yield systems of stars and planets that differ drastically. Slight tweaks in the starting positions of a few particles can change the number of planets, or even the number of stars.

Observations bear this out. Astronomers have long observed that stars often form binary systems. Recently the study of planets around stars other than our own Sun—exoplanets—has begun to flourish, and astronomers are observing wide variations in their sizes and in how they are distributed around their host stars.

Very slight tweaks to the early history of the solar system can make the difference between an asteroid that impacts Earth and kills off the dinosaurs and one that misses.

Thus, while a few ingredients, a few laws, and a strangely simple origin govern the broad framework and overarching flow of cosmic history, they are powerless to predict its rich local details. The world is like a tree that, following simple rules of growth, sprouts many branches, each different in detail, providing suitable homes for different birds and insects.

It is no contradiction that the history of, say, Sweden is more complicated than the history of the universe. Indeed, our fundamentals predict it.

THE FUTURE OF COSMIC COMPLEXITY

Heat Death and Its Remedies

The long-range future of the universe, on the face of it, looks bleak. Galaxies will keep receding from one another, stars will run out of nuclear fuel, the microwave background radiation will redshift into radio waves and peter out. Even before the emergence of big bang cosmology and the expanding universe, cosmologists worried about the "heat death" of the universe, as its approach to some sort of equilibrium seemed inevitable, after which nothing interesting would happen.

The first thing to say about this is that it's not an immediate worry. Our Sun has at least a couple of billion good years ahead of it, and stars continue to be born elsewhere in our galaxy, many of which (the M stars) will provide reliable heat for much longer than the Sun ever did.

With that much lead time, we should not underestimate how resourceful engineers might respond creatively. Dyson spheres around artificially constructed stars, together with energy-conserving technologies, could support intelligent life well beyond the natural lifetime of stars.

Especially good news here is that minds can run on very little energy—or maybe none. Quantum computers operate best in the cold and dark, where there's nothing fouling up their delicate works. A sufficiently complicated time crystal* of this kind could run through an elaborate program over and over again, giving joy to the AIs it contains.

Finally, we should remember that our scientific understanding of the universe is incomplete and evolving. The best thinking about every one of our fundamentals has changed drastically within just the past hundred years. Could we find ways to burn "dead" stars further, releasing the true bulk of their energy—the $E = mc^2$ energy of the nuclei they contain—in a usable form?† Could we re-create something like the big bang itself, giving birth to a baby universe? Could we tap into "dark matter" as an energy source?‡ We don't really know, and, of course, other pleasant surprises might arise. In the history of science and technology, a few billion years is a long time.

Complexity Within Simplicity

> The universe (which others call the Library) is composed of an indefinite and perhaps infinite number of hexagonal galleries, with vast air shafts between, surrounded by very low railings.
>
> —*Jorge Luis Borges*

---

* Time crystals are physical systems that spontaneously settle into stable loops of behavior. I proposed this concept in 2012, and many interesting examples have been discovered since then, both theoretically and experimentally.

† Unified theories suggest that protons are unstable, as we've discussed earlier. They also suggest the existence of excellent "catalysts" for proton decay—the so-called magnetic monopoles, or possibly cosmic strings. Thus, this speculation is not entirely groundless.

‡ Axions can be burned, in principle, but the energies you can generate this way are pathetic by solar standards, so this seems to be a last-ditch option.

Here, in sixteen words, I will supply a simple algorithm for producing the complete works of Shakespeare, at least one proof of Fermat's Last Theorem, and the paper that will win the Nobel Prize for physics in 2025:

1. Choose an ASCII character—a letter, number, space, or punctuation mark—at random.
2. Record it.
3. Repeat.

The output will contain all those promised things, and (much) more.

Borges's "The Library of Babel" expresses similar thoughts more poetically. Our program will generate "The Library of Babel," too.

This outrageous thought experiment illustrates how a very simple—that is, easily described—structure can contain vast complexities within it.

Our thought experiment might reflect reality. Quantum-mechanical wave functions contain vast amounts of information. The wave function for something as large as our universe could house the Library of Babel comfortably. Simple rules can generate capacious wave functions, just as our simple algorithm generates a capacious output.

Putting these thoughts together, it becomes tempting to think that the wave function of the universe is generated by a simple rule, yet to be discovered. If so, then the universe we experience and are part of is the ultimate fulfillment of "complexity emerges."

8

THERE'S PLENTY MORE TO SEE

> When I was a child, I spoke and thought and reasoned as a child. But when I grew up I put away childish things. Now we see things imperfectly, like puzzling reflections in a mirror, but then we will see everything with perfect clarity. All that I know now is partial and incomplete, but then I will know everything completely.
>
> —*Saint Paul, 1 Corinthians*

Visionaries of many persuasions have long suspected that there's much more to the world than our unaided senses reveal.

Saint Paul, in the passage above, contrasts the world that children construct, which takes things at face value, with the vague intuitions of thoughtful adults that there's more to be seen, and that we're headed toward a hoped-for and dazzling truth.

In Plato's Allegory of the Cave, Socrates describes a strange prison to his friend Glaucon. The prisoners inhabit a dark cave, and the only sights they are allowed to see are puppet shows projected on a wall. The prisoners mistakenly believe that what they see is the fullness of reality. Glaucon remarks, "This is an unusual picture that you are presenting here, and these are unusual prisoners," to which Socrates replies, "They are very much like us humans."

And William Blake, in a passage from *The Marriage of Heaven and Hell*, declared his faith that "if the doors of perception were cleansed, every thing would appear to man as it is: Infinite."

Science, in its account of the physical world, provides an inventory of what might possibly be observed. That inventory supports the visionaries. It reveals how impoverished natural human perception is, compared with the full content of physical reality. Science can help us to overcome our deficits. Much has been achieved, but much more can be done.

OPENING THE DOORS OF OUR PERCEPTION

Many animals inhabit a distinct sensory universe from humans. We share the physical world with them, but we experience it quite differently, not only at the level of intellect, but even at the level of raw perception.

Dogs and many other mammals live in a parallel universe dominated by smells. Dogs' noses are chemical laboratories, confronting incoming molecules with three hundred million receptors, compared to six million for humans. And a large portion—about 20 percent—of a dog's brain processes the result, compared with less than 1 percent for humans.

Bats navigate in the absence of light by sending out extremely high-pitched sounds—ultrasound—and analyzing the (ultra)sounds that bounce back. Human ears are deaf to ultrasound. They cannot be used for fine navigation, because the wavelength of humanly audible sound is too large. People have a poor sense of where the sounds they hear originate, in general.

Spiders construct sensory nets of another kind. Their webs are not only traps, but signaling devices, whose vibrations indicate the presence and position of prey.

Vision is our main portal to the external world, considering both how much information it gathers and how much of our brain—anywhere from 20 percent to 50 percent, depending on how you count*—is devoted to processing that information. Yet even here, our sampling of the external world is paltry relative to what's out there. Human vision samples the state of the electromagnetic field. But it samples only the radiation that happens to impinge on our pupils. Further, it is sensitive only to radiation within a narrow range of wavelengths, from about 350 to 700 nanometers (that is, around half a millionth of a meter, or a few hundred-thousandths of an inch). This defines "visible light." We don't take in a proper spectrum, either, even within that range of wavelengths. Instead, we have three† different kinds of cone cells, broadly tuned to different wavelength ranges, in-

* It is ambiguous because there are areas of the brain where information from several senses gets integrated, for example.

† There are several kinds of anomalous color perception, misnamed "color blindness," that are not terribly rare. Around 95 percent of humans have similar color vision, based on three types of cone cells that vary little between individuals. There are theoretical reasons, based on genetics, to believe that a significant fraction of humans, specifically mothers and full sisters of males with the most common color anomaly, have four kinds of cone cells. These "tetrachromats" might have super-normal color vision. But as far as I know, direct evidence for this is surprisingly scant.

volved in color vision, plus rod cells, also broadly tuned, that kick in for peripheral and night vision. Many snakes and other reptiles are sensitive to infrared. Bees are sensitive to ultraviolet, as are many birds. Birds also do a better spectral analysis of visible light. Their receptor cells contain oil droplets that selectively filter different wavelength ranges. Strangely enough, the order* of crustaceans known as mantis shrimps seem to be the best natural spectroscopists, by far. Depending on the species, mantis shrimps have between twelve and sixteen different kinds of receptors, compared with the human four. Their sensitivity extends well into the infrared and ultraviolet, and they are also sensitive to polarization (which humans are not).

Our ancestors inhabited a distinct sensory universe, too. It is difficult to imagine a world without eyeglasses, mirrors, magnifying lenses (and their advanced forms—microscopes and telescopes), artificial lighting and flashlights, clocks and watches, smoke alarms, thermometers, barometers, and a host of other devices that enrich our perception in many directions. Yet that was the world in which humans lived over most of our history.

Technology has already given us superpowers, and there is no end in sight. Receivers and generators of electromagnetic radiation both within and beyond the visible are becoming small and cheap, as are magnetic field sensors, generators and receivers of ultrasound, and devices that can perform chemical sampling of many kinds ("artificial noses"). The doors of perception are opening wider, as part of everyday life.

---

* That is, a diverse range of related species. There are more than 450 recognized species of mantis shrimp.

HARD-EARNED REVELATIONS

Other projects for expanding our perception call on extraordinary resources from many parts of science and technology. They are meant to address grand questions by consulting Nature in new ways. The new perceptions they yield will not for the foreseeable future be part of everyday life. But people have been inspired to work hard to answer them, simply because they are interesting.

Here I will briefly describe two big projects that have expanded our perception of the world in recent years. They are examples of *planned discovery*, where we put sharply posed questions to Nature and expect to get answers. In each case, I'll explain why we're asking the question, what it makes us want to explore, and how we take on the exploration.

These projects push the limits of what we know how to do,* in order to expand the horizon of knowledge. Thus, they give stress tests to our fundamental understanding.

THE HIGGS PARTICLE

Why We Look, and What We Look For

Imagine a planet or moon encrusted with ice, beneath which a vast ocean lies—a planet like Jupiter's Europa. Imagine that

* That is, what we *think* we know how to do.

within that ocean a brilliant species of fish evolves—a species so intelligent that they take up the physics of motion. Because the way that bodies move in water is complicated, their work produces many interesting observations and rules of thumb, but no coherent system. And so it goes, until one day a genius among fish, we'll call her Fish Newton, has a startling new idea. Fish Newton proposes new, much simpler laws of motion—Newton's laws. They are much simpler than the old rules, but they don't describe the way that things actually move (in water, that is). Fish Newton proposes that you can reproduce the observed motions from the new, simpler laws if you assume that a medium fills space. Her hypothetical medium—the material we call water—affects the behavior of bodies. Fish Newton's idea reconciles the complications of observed reality with a more fundamental, underlying simplicity.

Ah Love! could thou and I with Fate conspire
To grasp the sorry Scheme of Things entire,
Would not we shatter it to bits—and then
Remould it nearer to the Heart's Desire!

—Omar Khayyam (Fitzgerald translation)

When the appearance of things is disappointing or discordant, we can, like Fish Newton, imagine a better world, and then try to build up ours within it. This was the strategy that led to the modern understanding of the weak force.

The medium that complicates the weak force is called the Higgs condensate, after Peter Higgs, a Scottish physicist who made important contributions to its theory.* It was first introduced theoretically, as a way to get more beautiful equations, à la Fish Newton.

Once we strip away the Higgs condensate, we can construct a theory of the weak force that looks a lot like our theories of the strong and electromagnetic forces. In the imaginary world, the weak force is mediated by gluon-like (and photon-like) particles—the *W* and *Z* bosons—that change and respond to two new kinds of charge. The new kinds of charge—let's call them weak charge *A* and weak charge *B*—are similar to, but distinct from, the three color charges of quantum chromodynamics and the one electric charge of quantum electrodynamics. The weak force—and only the weak force—can transform a unit of type *A* charge into a unit of type *B* charge, or vice versa. Since particles are defined by properties, these transformations of weak charge change one kind of particle into another. This, in our deeper understanding, is the nature of the weak force's transformative power.

The reason we need to bring in the Higgs condensate is that in the world we observe, the *W* and *Z* bosons, unlike gluons or photons, have nonvanishing mass. To consummate the analogy and get similarly beautiful equations, we must bring in a medium, to slow them down.

This medium-based theory of the weak force took shape over the 1960s. Over the 1970s, experimental evidence in its favor began to accumulate, and eventually it got overwhelming.

* As did several others. This is not the place, nor am I the scholar, to present the complicated history surrounding the genesis of the theory.

But one big question remained unanswered: What is that all-important, ubiquitous medium—the Higgs condensate—made out of?

People gave many speculative answers to that question. Some postulated that it is made out of several different particles, and they invoked new forces or even new dimensions of space. But the simplest, most radically conservative possibility was to make it from a single new particle—the Higgs particle. It became important to check whether Nature uses this simplest option.

How We See It

If the Higgs condensate is made from just one ingredient, then we can say a lot about that ingredient. Roughly speaking, if the Higgs particle is a chunk of the condensate, the only question is how big a chunk. Thus, all the properties and behaviors of the Higgs particle can be predicted, once you know its mass. This welcome specificity meant that experimenters could plan their Higgs-hunting strategy with quite definite ideas about what they were looking for, and how they would recognize it if they found it.

In order to "discover the Higgs particle," you must do two things: You must produce some of them and you must get evidence of their fleeting existence. Both steps are challenging. To produce heavy elementary particles, you must concentrate a lot of energy into a very small volume. This is done at high-energy accelerators, where beams of rapidly moving protons (or other particles*) are made to collide with target materials, or with one another. In the years prior to 2012, Higgs particle searches were mounted with a succession of ever-higher energy concentrations, but they came up empty. We know now, in retrospect, that they simply didn't bring in enough energy. The Large Hadron Collider, or LHC, finally did.

The home of the LHC is a circular underground tunnel measuring about twenty-seven kilometers (seventeen miles) around, beneath a rural area straddling France and Switzerland. When the LHC is operating, two narrow beams of protons traverse the tunnel in opposite directions within a pipe that threads it. Moving at nearly the speed of light, the protons make eleven thousand orbits per second.

At four points the beams cross. Only a small fraction of the protons collide, but this still amounts to nearly a billion collisions per second. All that firepower produces the concentrations of energy it takes to make Higgs particles.

The next task is to detect them. Enormous, densely instrumented detectors surround the crossing points. One of them, the ATLAS detector, is more than twice as large as the Parthenon. The detectors track the energies, charges, and masses of the particles that emerge from the collisions, as well as their directions of motion. They feed all this information, at the rate of 25 million gigabytes per year, to a worldwide grid that links thousands of supercomputers.

All that information gathering is necessary because:

* Beams of electrons, antielectrons, antiprotons, photons, and various atomic nuclei, and even neutrinos and antineutrinos, have all been used at high-energy accelerators, for different kinds of experiments. The discovery of the Higgs particle was done using two colliding beams of protons.

- The events are complicated. Typically, ten or more particles stream out from each one.
- Few of the events—less than one in a billion—ever contained Higgs particles.
- Those events that do contain them, don't contain them for long. The lifetime of a Higgs particle is about 10^{-22} seconds, or a tenth of a trillionth of a billionth of a second.
- The rare events that briefly contained Higgs particles also contain a lot of other stuff.

In short, if you're going to find the Higgs particle, you have to understand and monitor the rest of what's happening very well, indeed—*and* you've got to latch on to some nearly unmistakable consequence of a Higgs particle's brief existence. Otherwise you'll get inundated with false positives.

The discovery of the Higgs particle was announced on July 4, 2012. The signal was an excess of high-energy photon pairs. Such pairs were predicted to arise from Higgs particle decays, and the excess swamped any other plausible source.* Since then, several other signals, arising out of other ways that Higgs particles can decay, have been detected as well. So far, the rates at which all these signals have occurred agree with theoretical predictions.

In "seeing" the Higgs particle, we humans expanded our perception. We peered into a behavior that Nature reveals only

* Many photon pairs are produced by other processes, but only pairs with a distinctive amount of energy and momentum can be ascribed to Higgs decays. By comparing the rate of production for pairs with and without the favored amount of energy and momentum—on-resonance and off-resonance, we say—you define "the excess."

rarely, and for very short times, and only after vigorous prodding. To perceptive human minds, empty space will never look empty again. Fish Newton, and Peter Higgs, nailed it.

GRAVITATIONAL WAVES

Why We Look, and What We Look For

Let's recall that John Wheeler, the poet of general relativity, summed it up this way: "Space-time tells matter how to move; matter tells space-time how to bend." Wheeler's summary is catchy, but it is misleading, or at least incomplete, without this important addition: *Space-time is a form of matter, too.*

Specifically, it is wrong to think that the curvature of space-time is entirely dictated by something else—that is, "matter." Bending space-time requires energy, and energy causes space-time to bend. In this way, curvature participates in its own creation. Space-time, in short, has a life of its own.

We've heard this song before. The crowning triumph of Faraday's field concept, and more specifically of the Maxwell equations that express his concept mathematically, was the discovery of electromagnetic waves. In such waves, the electromagnetic field takes on a life of its own. Changing electric fields create changing magnetic fields, which create changing electric fields, and so on, indefinitely. A self-sustaining disturbance in the fields moves through space. If the disturbance repeats at an appropriate wavelength, we will see it as light.

We've also learned to "see" other wavelengths, using detectors designed for the purpose, such as radio receivers or microwave dishes.

In a similar way, Einstein's curvature field, which encodes gravity, can also support self-sustaining disturbances. These are called gravitational waves. In gravitational waves, bending of space-time in some directions causes bending in others.

The equations for gravitational waves very much resemble those that govern electromagnetic waves—with different interpretations of the symbols, of course.* The kinds of sources that trigger the waves are different: Moving electric charges radiate electromagnetic waves, while moving masses radiate gravitational waves.

Despite their *qualitative* resemblance, there is a big *quantitative* difference between electromagnetic and gravitational waves. This quantitative difference arises because, according to general relativity, space-time is extremely stiff. Because of this stiffness, even rapid motions involving large amounts of mass produce only tiny wiggles in space-time. That is both good news and bad news.

The good news is that when we detect gravitational waves, they bear messages from some of the wildest, most interesting events in the universe, in which big things go whipping about. Gravitational waves give us a new way to perceive the universe that is especially attuned to such events.

The Laser Interferometer Gravitational-Wave Observatory, or LIGO, was designed with a few spectacular sources in mind.

* One feature they share is that gravitational waves travel at the speed of light.

These include blasts from the moments when systems of two black holes, or two neutron stars, or one of each, that have been orbiting each other at last spiral in and merge. As they lose energy to gravitational radiation, the orbits of those systems decay. The decay is slow and gradual until the last few moments, when things move especially fast. It is only then that detectable bursts of radiation are produced.

The bad news is that gravitational waves are difficult to detect.

How We See It

The basic concept that eventually matured into LIGO was contained in a paper published by Rainer Weiss in 1967. To reach the sensitivity necessary to detect gravitational waves, many technological innovations were required. The first successful observation of gravitational waves came almost fifty years later. Weiss, together with Kip Thorne and Barry Barish, received the Nobel Prize in 2017 for their work on LIGO.

To envision how LIGO detected gravitational waves, imagine three objects at the vertices of a big (imaginary) L. To keep things simple, let's assume that they're floating in space. As a gravitational wave passes, space itself is distorted, so that the distances among the three objects change with time. If we have a way to compare the lengths of the L's arms, we can look for that effect. It provides a signal for gravitational waves.

Some rough calculations, however, give discouraging estimates for the size of the effect. The fractional change in the lengths is 10^{-21}, or a part in a billion of a trillionth. This seemed, to most physicists, impossible to detect. But Rainer Weiss and friends brought in new ideas and tricks. They used mirrors for their reference objects. They kept the mirrors far apart* and bounced light beams back and forth many times across each arm. This repeated passage of light in effect magnified the lengths of the arms. A standard technique—interferometry—allows you to compare the lengths of light paths to within a fraction of a wavelength. Putting it all together, the tiny ratio of light's wavelength to the magnified arm length can get you to 10^{-21}.

These tricks build you a detector that is exquisitely sensitive to the relative motions of the mirrors. The next challenge is to separate motion caused by gravitational waves from all the other things that might change the inter-mirror distances.

There are lots of things to worry about, of course. The planning documents and discovery papers of the LIGO group go into great depth and detail about the precautions they take and the consistency checks they perform. Here I will mention only one of the most serious. Vibrations of the Earth on which the experiment rests, due to anything from low-grade earthquakes to bad weather to passing trucks, are unavoidable. To suppress the effects of such vibrations, the mirrors are suspended on a quadruple pendulum and stabilized by active feedback. These are marvels of engineering, which take the art of shock absorption and noise cancellation to new levels.

On the other side of the ledger, vibrations due to gravitational waves are predicted to have some special characteristics, which aid in positive identification. The most basic is that they

---

* Several kilometers, eventually.

must excite two separate detectors, in two locations, with matching but offset patterns of motion that are consistent with a disturbance that travels at the speed of light. In more detail, the theory of black hole and neutron star mergers predicts how the vibrations should look, as a function of time, if they are caused by gravitational waves from those sources.

The first successful detection of gravitational waves took place on September 18, 2015. This detection matched predictions for the burst of radiation from a merger of two black holes, with masses roughly twenty to thirty times that of our Sun, about 1.3 billion light-years away.

Since then about fifty more events have been detected. An especially interesting one occurred on August 17, 2017. This matched predictions for the merger of two neutron stars. Alerted to this event, astronomers also observed it in several parts of the electromagnetic spectrum, including a gamma ray burst and a lingering visible afterglow. This inaugurated a new kind of "multi-messenger" astronomy, which promises to enrich our perception of strange, faraway events.

THE FUTURE OF PERCEPTION

Distributed Sensoria

> listen: there's a hell
> of a good universe next door; let's go
>
> —*e. e. cummings*

The "phantom hand" illusion is a startling experience. In it, you hide your right hand behind a partition and look at a fake rubber hand near it. A friend taps and strokes both your unseen real hand and its visible facsimile in a random but synchronous way. After a brief interval—typically less than a minute—you will experience the taps and strokes as originating from the rubber hand rather than your own. Diane Rogers-Ramachandran and Vilayanur Ramachandran, pioneers in the study of this and related illusions, called attention to its profound implications:

> All of us go through life making certain assumptions about our existence. . . . But one premise that seems to be beyond question is that you are anchored in your body. Yet given a few seconds of the right kind of stimulation, even this axiomatic foundation of your being is temporarily forsaken.

A few years ago, for an hour or so, I was in two places at once. I was sitting at home in Cambridge, Massachusetts, and at the same time attending a conference in Gothenburg, Sweden. I got that way through a full-body version of the phantom hand illusion. I saw and heard the world through the "eyes" and "ears" of a robot whose gaze and attention I controlled remotely, using a joystick. I could also "walk around" and talk with people, while they saw my facial expressions displayed on a screen that formed part of robotic me. I gave a short talk, pacing the stage and picking up on the audience's reactions, joined in a panel discussion, and mingled at coffee breaks.

At first, as I was learning how to navigate the system, I was acutely aware of the artificiality of the situation. But after a half hour or so, as the mechanics became second nature and no longer required conscious direction, I felt as if I really were in Gothenburg. Yet I remained aware, at the back of my mind, that I was also in Cambridge, sitting in front of a computer screen. My consciousness had expanded—my robot had extended my self.

The system I was using was crude. No one would mistake the ProBeam platform for a human body, any more than they'd mistake a rubber hand for flesh and blood. Yet it led me to a compelling experience. In the future, more richly endowed platforms and, at the other end, more immersive virtual reality feedback will support sensoria that are widely distributed in space yet deeply integrated within our minds.

Quantum Perception and Self-Perception

> I think I can safely say that nobody understands quantum mechanics.
>
> —*Richard Feynman*

> I consider that I understand an equation when I can predict the properties of its solutions, without actually solving it.
>
> —*Paul Dirac*

Natural human perception is a poor fit to quantum mechanics. In the quantum world, many possible arrangements and behaviors coexist. If you look, you'll see just one of them—and you can't tell in advance which one. No single set of perceptions (that is, observations) can do full justice to the state of a quantum system.*

The crowning achievement of natural human perception, by contrast, is to give us a representation of the world in terms of objects with more or less predictable properties occupying more or less definite positions in three-dimensional space. That's very useful information for navigating everyday life, and we extract it effortlessly. But fundamental understanding reveals that there's plenty more to see, and quantum mechanics takes it to another level.

Fortunately, there are ways, as yet little explored, that we can retrofit the quantum world to human perception. If we can compute an interesting state—say, the state of the quarks and gluons in a proton, or of the electrons and nuclei in a molecule, or of the qubits in a quantum computer—then we can also compute how our observations of those things would have turned out, as many times as we like, *as if we had made them*. Then we can present the results as "normal" perceptions, on many displays, all presented in parallel. In this way, physicists, chemists, and tourists could immerse themselves in the quantum world, and maybe finally come to understand it.

Know thyself.

—Inscription at the temple of Apollo, Delphi

An oddly parallel issue arises in our self-perception. Many things are happening simultaneously within our brains, but our

* There's more on this in our concluding chapter.

natural consciousness only allows us to attend to one at a time, and much is hidden from it altogether. You can switch attention from one working module to another, but it's difficult and unnatural to focus simultaneously on more than one.* As our ability to monitor and interpret brain states improve, it will be possible to present our inner selves to our perceiving self through our visual system, on displays, bypassing the filter of natural consciousness. More will come through, and less will be hidden. People will come to know themselves, and perhaps others, in new ways, and more deeply.

* I'm aware that these two sentences are a crude description of a very complex reality. They are correct in spirit, and sufficient for me to make my point.

9

MYSTERIES REMAIN

> The most beautiful thing we can experience is the mysterious. It is the source of all true art and science. He to whom the emotion is a stranger, who can no longer pause to wonder and stand wrapped in awe, is as good as dead—his eyes are closed.
>
> —*Albert Einstein*

Although we understand *a lot* about how the world works, there are still big mysteries. These three great questions came up earlier:

- What triggered the big bang? Could it happen again?
- Are there meaningful patterns hidden in the apparent sprawl of fundamental particles and forces?
- How, concretely, does mind emerge from matter? (Or does it?)

Here we will focus our exploration on two big mysteries that are more sharply focused. They are at the cutting edge of research aimed at deepening our fundamental understanding of the physical world. The first mystery surrounds a strange feature of fundamental laws. They work almost, but not quite exactly, the same if you run time backward. The second mystery emerged from a confounding discovery. Astronomers have encountered, in a variety of situations, what appear to be gravitational forces that have no visible cause. Their observations, on the face of it, seem to indicate the existence of a "dark side," consisting of two new forms of matter, "dark matter" and "dark energy," which have somehow escaped previous notice, despite providing most of the mass in the universe.

A promising idea has emerged that may help to solve these mysteries. The time-reversal problem has led many physicists to suspect the existence of a new kind of particle, the *axion.* The lingering afterglow of axions, left over from the big bang, has the right properties to be dark matter. A flurry of developments surrounding this idea have led to a spirited race for discovery, involving hundreds of scientists around the world.

TIME REVERSAL (T)

Time's Mirror Image

Few aspects of experienced reality are as obvious as the asymmetry between past and future. We remember the past, but can only guess about the future. If you run a movie—say, Charlie Chaplin's *City Lights*—backward, it doesn't look remotely like a sequence of events that could unfold in reality. You would never confuse it with a legitimate movie.

Yet beginning with the birth of modern science, in Newton's classical mechanics, and until quite recently, the fundamental laws had the character that you could run them backward in time. That is, the laws you need to predict past states, given present states, are the same laws that you use to predict future states. For example, if you imagine filming a movie of planets orbiting the Sun, according to Newton's laws, and run it backward, the movie will still obey Newton's laws. This feature of the laws is called time-reversal symmetry, or T for short.

Time-reversal symmetry continued to hold up as the scope of the laws expanded. Maxwell's equations of electromagnetism and Einstein's revised equations of gravity both have it, for example, as do the quantum versions of those equations. And observations of fundamental interactions seemed to bear T out.

This contrast between everyday experience and the fundamental laws poses two problems. One is how the actual universe finds a preferred direction for the flow of time. We got an answer to that in chapters 6 and (especially) 7, where we saw that gravity started way out of equilibrium.* The other is, simply, Why? Why in the world should our fundamental description of Nature have this feature, T, that the world we experience so blatantly lacks?

* Of course, why *that* happened is an obvious follow-up question. We discussed some relevant ideas, specifically inflation and complexity within simplicity, in chapters 6 and 7.

Why? First Pass: Rock Bottom

Parents of young children sometimes have the exasperating experience of never-ending "Why"s. (*Why* do I have to go to bed? Because people need to rest. *Why?* Because their bodies get tired. *Why?* Because after we use our muscles for a while they don't work as well. *Why?* Because they use up the food we ate, and some junk gets left behind, which has to get cleaned up. *Why?* Because everything runs down, according to the second law of thermodynamics. *Why?* Because during the big bang, gravity was out of equilibrium . . .) Eventually you will run out of answers.* At some point you hit rock bottom, with some answer so basic that it can't be further explained: That's just the way it is.

It was unclear, while T appeared to be an exact feature of fundamental laws, that asking "Why?" would be fruitful. It appeared to be an elegant, if slightly peculiar, property of the laws. T might be rock bottom. Most physicists thought that it was.

Why? Second Pass: Sacred Principles

The situation changed in 1964, when James Cronin, Val Fitch, and their collaborators discovered a tiny, obscure effect in the decays of K mesons† that violates T. Since T is not quite right, it can't be rock bottom. At that point there clearly was a ques-

* Alternatively, your answers might put the kid to sleep.

† K mesons are highly unstable, strong interacting particles (hadrons) whose properties can be, and have been, studied in great detail at high-energy accelerators. They are the lightest hadrons that contain strange (*s*) quarks.

tion to pursue further: Why does Nature obey T very nearly, *but not exactly*? That question proves to be wonderfully fruitful.

In 1973, Makoto Kobayashi and Toshihide Maskawa made a theoretical breakthrough on this problem. They built upon the imposing framework of quantum field theory and our Core theories of the forces (which at the time were not yet firmly in place). That framework is very rigid, as I mentioned earlier—you can't easily change it without ruining its consistency. No one knows how to change its structure without violating the sacred principles* of relativity, quantum mechanics, and locality. But you can add to it. What Kobayashi and Maskawa discovered is that by adding a third family of quarks and leptons† to the two that were then known, you have the possibility of introducing an interaction that violates T and generates the effect that Cronin and Fitch observed. With only the two known families, there was no such possibility.

Soon after Kobayashi and Maskawa's work, the particles from the third family they predicted began to show up at particle accelerators, as they operated at higher energies. Since then, many experiments have vindicated the interaction they proposed as well.

That isn't the end of the story, though. Besides the interaction that Kobayashi and Maskawa made use of, there is exactly one *other* possible interaction that violates T but is completely consistent with the rigid framework of our Core theories and

---

* Of course, no scientific principles are sacred in a dogmatic, theological sense. But if relativity, quantum mechanics, or locality is wrong, we've got a lot of unlearning to do, because those principles work well and explain a lot. In other words, they're probably closer to rock bottom than T is.

† For more on these "bonus" particles, see the appendix. The details are not crucial for what follows.

quantum field theory. This interaction isn't necessary to explain what Cronin and Fitch saw, or any other observation. Nature doesn't seem to use it. *Why?*

Why? Third Pass: Evolution

In 1977, Roberto Peccei and Helen Quinn proposed an answer to that third and potentially final "Why?" about T. It is a theory of evolution, opened up by expanding the Core. What they proposed is that the strength of the unwanted additional interaction is not simply a number, but a quantum field, which can vary in space and time. They showed that if the new field has some appropriate, reasonably simple properties, then the forces acting upon it will tend to drive it toward zero. Peccei and Quinn implicitly assumed that the field takes on its favored value, zero. Big bang cosmology suggests that the field evolves toward that value.*

That would leave us, at last, with a satisfying answer to our questions: T is very nearly, but not exactly, a feature of fundamental laws, as an indirect consequence of how deeper principles—relativity, quantum mechanics, and locality—act upon the fundamental ingredients of the world.

These theoretical ideas have a dramatic consequence. We'll take it up shortly. First, let's visit the dark side.

---

* The difference could supply the dark matter of the universe, as we'll soon discuss.

THE DARK SIDE

Dark matter and dark energy have a similar character, so it makes sense to introduce them together. They both refer to observed motions that have no apparent cause. It would be more accurate, if less evocative, to say we have "unexplained accelerations," rather than "dark matter" and "dark energy." But the extra motions are all of a pattern, which suggests that they are caused by gravity from sources that are otherwise invisible. In order to account for all the observations, we need two distinct new sources. These, by definition, are dark matter and dark energy. Let me emphasize that neither dark matter nor dark energy is "dark" in the usual sense of English. Both have proved to be invisible, so far. Neither emission nor absorption of light has been detected from where the "dark" stuff is supposed to be.

Dark matter could be composed of a new kind of particle, produced during the big bang, that interacts only very feebly with ordinary matter. Dark energy could be a universal density of space itself. Those are the most popular ideas about what they are among researchers on the subject, and they do account for a wide range of observations fairly convincingly. Other ideas have advocates, too, but they're (even more) speculative.

Problems similar to this—missing acceleration problems—have happened before in astronomy. A little history will set the stage for us.

For many decades following their introduction in 1687, Newtonian mechanics and his law of gravitation—what he called his

"System of the World"—went from triumph to triumph. Many people made much more accurate observations of astronomical motions, and others made much more accurate and extensive calculations of the theory's predictions. Almost without exception, the observations were consistent with the predictions.

There were, however, two nagging problems. They concerned the motions of the planets Uranus and Mercury. Clear discrepancies emerged between the predictions of Newton's theory and the observed positions of those planets. The discrepancies were quite small—they amount to far less than the size of the Moon in the sky, for example—but they were well outside what the accuracy of the observations could permit. Something had to give. Either the calculations were missing something, or the theory was wrong.

When an otherwise extremely successful theory hits a snag, the conservative hypothesis is that something is missing. And so both John Couch Adams and Urbain Le Verrier considered the possibility that there might be another planet, not yet recognized, whose gravity was throwing Uranus off course. In other words, they proposed that a very specific kind of "dark matter" was involved.

Adams and Le Verrier calculated where the new planet would have to be, and where it would appear in the night sky. Le Verrier communicated his prediction to the Berlin Observatory. The observers looked, and they saw it. The new planet, discovered in 1846, is what we now call Neptune.

Le Verrier tried a similar approach for the problem with Mercury. He postulated the existence of another new planet, which he called Vulcan. Vulcan had to be very close to the Sun, so that its gravity would influence Mercury but not make a noticeable impression on the other planets. That would also explain why Vulcan had not been observed, since the Sun presents a formidable background.

Astronomers set out to find Vulcan, especially during solar eclipses. Quite a few even reported success. But none of those sightings convinced the community, and the problem festered. Ultimately the solution came from quite a different direction. In 1915, Albert Einstein proposed a profoundly new theory of gravity, his general relativity theory. Although Newton's theory and general relativity are based on radically different ideas, in many situations they give similar predictions. Within the solar system, by far the biggest difference (still a small one) concerns the motion of Mercury. One of the first major triumphs of Einstein's theory, already in his original paper, was its ability to reproduce the observed motion of Mercury, without requiring an additional planet. After that, Vulcan was never seen again.

"Dark energy" is another theoretically motivated modification to the law of gravity, also considered by Einstein. He called it by a different name: the cosmological constant. It builds on general relativity. If you stay within the conceptual framework of general relativity, there is basically just one way to change the law of gravity—one "free parameter," we say—and that's to add a cosmological constant. At the time when he considered it, there were no observations that required a nonzero cosmological constant, and in the spirit of Occam's razor, Einstein set it to zero. But it was ready for use, if observations required it.

As a little joke, to summarize their historical parallels, we could say that dark matter is from Neptune, while dark energy is from Mercury. The encouraging message from history is that good scientific mysteries often find worthy solutions.

Dark Matter

The modern dark matter problem plays out over the whole universe. On several scales, in many different circumstances, astronomers observe "excess" acceleration. Here I'll mention two classes of observations, which encompass dozens if not hundreds of well-documented examples.

The first concerns the speed at which stars and gas clouds in the outer fringes of galaxies rotate around those galaxies. One of Kepler's laws, which today follows both from Newton's and Einstein's theories of gravity, connects the speed of rotation around an orbit to the amount of mass inside. Thus, from the observed rotation speeds, you can infer how mass is distributed within a galaxy of interest. What people find is that to explain the observed speeds, you need lots of mass in places where there isn't much light being emitted. It seems, in essentially all cases that have been studied, that the galaxy is surrounded by an extended halo of dark (invisible) matter. Indeed, it would be more appropriate to say that the lit-up part of the galaxy is an impurity within a cloud of dark matter. The dark matter halo, when you add it all up, weighs about six times more than the visible impurity.

The second concerns the bending of light, or what is called gravitational lensing. Astronomers have observed in many cases that the image of very distant galaxies is grossly distorted, as if you were looking at it through a glass of water or a Coke bottle. This occurs in particular when the light of the galaxy you're looking at passes through a region of space containing a cluster of other galaxies. General relativity predicts that gravity should bend light, so the existence of this gravitational lensing is not surprising. What is surprising is the size of the effect. Here again, astronomers find that they need the galaxies in the cluster to weigh about six times what the visible stars and gas clouds supply.

These and other observations suggest that dark matter provides about 25 percent of the mass in the universe. "Normal" matter—the kind that we understand and are made of—provides about 4 percent. Most of the rest is dark energy.

Dark Energy

A different class of observations leads us to dark energy. Here there is an important backstory. Albert Einstein formulated his theory of gravitation, general relativity, in 1915. Not long afterward, in 1917, he considered a modification of the equations, to allow for what he called a "cosmological constant." Physically, introducing the cosmological constant corresponds to assigning a nonzero density to space itself. Thus, a nonzero value of the cosmological constant means that every unit of volume in space contributes an equal, nonzero amount to the total mass of the universe, even when there's (apparently) nothing there.

A nonzero cosmological constant fits easily into the framework of general relativity. It does not require a significant change to the theory's basic principles. Matter still bends space-time in the same way as before, and matter responds to space-time curvature in the same way as before. The cosmological constant merely recognizes the possibility that space-time itself, a material that general relativity allows to bend, push, and shake, might also have inertia. Other possible modifications of general relativity are, by contrast, either highly contrived or tiny in their physical effects.

The cosmological constant's universal density comes paired with a peculiar partner property. Together with space's positive density of mass, one must include a negative pressure, whose magnitude is equal to the density times the square of the speed of light. That relation between density and pressure is the analogue, for mass tied up in space, of the more famous relation $E = mc^2$, which connects energy to mass for particles.

During the 1990s, the cosmological constant got rebranded as *dark energy.* The new name reflects a new attitude. Modern physicists, internalizing the lessons they learned in understanding the other forces, recognize that the density of space is not merely a parameter that appears in general relativity, whose value has no other meaning. It is tied up with the rest of physics, and its value can receive contributions from many different sources. In a universe filled with restless quantum fields, it would be surprising if space didn't have inertia.

In 1998, astronomers discovered dark energy. What they observed, to be specific, is that the rate of expansion of the universe has been increasing, consistent with a universal negative pressure. This was inferred from measurements of redshifts, in the spirit of Hubble, but using supernovas in place of

Cepheid variables. Supernovas, being much brighter, allow access to larger distances.

The density of space they measured is, by most standards, exceedingly small. A volume of space equal to the Earth's volume weighs about 7 milligrams. Within the solar system, or even the galaxy, the mass contributed by space is utterly negligible compared with the mass contributed by ordinary matter (or dark matter). But such is the vast emptiness of intergalactic space that this small density, present everywhere, comes to dominate the total mass of the universe.

Dark energy presently accounts for about 70 percent of the universe's mass. Nobody knows why several different, much larger contributions from various sources—some positive, some negative—conspire to give that particular final result. It's a big cosmic mystery.

A Cosmological "Standard Model"

Understanding that together dark matter and dark energy (hypothetically) presently constitute most of the mass in the universe, we might anticipate that they played a significant role in the history of the universe, too. To "run the movie backward" and check that intuition, we need to be more specific about what the properties of dark matter and dark energy are. Revisiting the big bang gives us a chance to learn about dark side properties. If we guess wrong about them, then our model of the big bang won't produce the universe we observe.

Given how little we know about the dark side, the task of

guessing how dark matter and dark energy might have behaved during the early moments of the big bang might seem hopeless. Fortunately, it turns out that we don't need to know much, and some simple guesses have worked out remarkably well.

For dark matter, we assume that it is made from some kind of particle that interacts feebly both with normal matter and with itself. We also assume that it was in equilibrium with the rest of the cosmic fireball early on, but that it cut away relatively shortly thereafter, becoming a lingering afterglow of the kind we discussed in chapter 6. One subtle point, on which some early proposals for dark matter foundered, is that when they cut away, the particles must have been moving much slower than the speed of light.* Because (by assumption) gravity is the only relevant force, and gravity doesn't distinguish among different forms of matter, that's all we need to know. We can calculate how dark matter moves, and how it affects the rest of the universe, once it has cut away. This defines the so-called cold dark matter model.

For dark energy, we adopt Einstein's idea that it represents a universal density of space itself, and that it is associated with a universal negative pressure.

Given those assumptions, we can run the density contrasts we observe in the cosmic microwave background radiation, which date from 380,000 years after the big bang, forward to the present. The addition of dark matter makes the instability work faster than it otherwise would. With dark matter, the model universe evolves to look like ours. Without dark matter,

* If the particles move too fast, they blur the growth of gravitational instabilities, and you get model universes that don't look like ours.

it doesn't. In this way, the dark side allows us to fulfill the promise of big bang cosmology to produce the structure we observe in the universe today, starting from tiny seed density contrasts, through gravitational instability.

AXIONS: QUANTA THAT CLEANSE

When I was a teenager, I sometimes accompanied my mother to the supermarket. On one of those trips, I noticed a laundry detergent called Axion. It occurred to me that "axion" would be a good name for an elementary particle. It was short, catchy, and would fit in nicely alongside proton, neutron, electron, and pion. I had the passing thought that if I ever got a chance to name a particle I'd call it the axion.

In 1978, I got my chance. I realized that the Peccei-Quinn idea, to introduce a new quantum field, had an important consequence that they hadn't noticed.* Quantum fields produce particles—their quanta—as we discussed earlier. And this particular field produced an extraordinarily interesting particle. The new particle had the intriguing technical feature that it cleaned up a problem with an axial current. The stars were aligned, and axions entered the world—or at least the world of physics literature.

(By the way, the naming would never have got past the editors of *Physical Review Letters*, or possibly the makers of Axion

* Steven Weinberg had the same realization, independently.

detergent, if I'd broadcast my true motivation prior to publication. Instead, I mentioned the axial current.)

Looking for Their Lingering Glow

Axions have the right properties to provide the cosmological dark matter. They interact very feebly with normal matter and with each other. They get produced at a high temperature and then later break free from the cosmic fireball. Their lingering afterglow, the axion background, fills the universe. The calculated density of the axion background is consistent with the observed density of dark matter, and axions are produced almost at rest. Thus, the axion background fulfills the assumptions of "cold dark matter" cosmology.

It's a beautiful story, but is it true? Axions, as we've said, interact only feebly with matter—but the theory tells us that they do interact, and how. In order to detect the axion background, we'll need to design sensitive new kinds of detectors, tailored to their properties. Hundreds of physicists, both theoreticians and experimentalists, are taking up this challenge today. If there's justice in the world, and luck, we may soon witness a success story worthy of a place beside the discoveries of Neptune, the cosmic microwave background, the Higgs particle, gravitational waves, and exoplanets. Scientific mystery stories often have worthy solutions.

THE FUTURE OF MYSTERY

How Mysteries End

Val Fitch, the hero of T violation who appeared earlier, was a wise man with a subtle sense of humor. He was chairman of the Princeton University physics department when I was a professor there, early in my career. In telling him about my emerging ideas on axions and dark matter,* I spoke about T violation as if it were an established fact from ancient history. After all, I'd never known anything else. At some point, he smiled gently and said, "Yesterday's sensation is today's calibration."

That is the fate of successful scientific mystery stories. I lived through a similar process, on the receiving end, with asymptotic freedom and QCD (quantum chromodynamics). For several years after our breakthrough, there was a lot of excitement and doubt around the question of whether it really did solve the mystery of the strong force. Big international conferences featured talks on "Tests of QCD," which reported progress in using the theory to make predictions and in testing it experimentally. Gradually, though, excitement dwindled, as the doubts faded. Today, the same sort of work, now vastly more sophisticated, goes on behind the scenes. It is called "calculating background." Yesterday's sensation is today's calibration, and tomorrow's background.

* And also the cosmic asymmetry between matter and antimatter.

Knowing and Wondering

Besides the future of particular mysteries, there are interesting questions around the future of mystery itself.

The Clay Foundation has offered a prize of one million dollars for a proof that QCD predicts that quarks are confined. Physicists have lower—or I'd rather say different—standards. As far as I'm concerned, we've moved way beyond proving that quarks are confined. With the help of our silicon friends, we can calculate what kinds of particles QCD produces, with no serious room for error. Isolated quarks are not among them. Indeed, the calculations give us the particles with the masses and properties of the particles we observe in Nature—no more, and no less.

Should a supercomputer get the prize? Or should its programmers?

In 2017, AlphaZero, a highly innovative computer program using artificial neural nets, after being given the rules of chess, played games against itself for a few hours, learned from the experience, and achieved superhuman performance. Does AlphaZero understand chess? If you're tempted to answer "No," I refer you to Emanuel Lasker, the world chess champion for many years, from 1894 through 1921.*

> On the chessboard lies and hypocrisy do not survive long. The creative combination lays bare the presumption of lies; the merciless fact, culminating in checkmate, contradicts the hypocrites.

---

* Lasker also did important work in pure mathematics.

Examples like these show that there are ways of knowing that are not available to human consciousness. But really, this should not come as fresh news. Humans themselves know many things that are not available to human consciousness, such as how to process visual information at incredible speeds, or how to make their bodies stay upright, walk, and run.

The genomes of humans and of Earth's other creatures are another great repository of unconscious knowledge. They have solved many complex problems that arise in building up organisms that flourish, accomplishing feats far beyond the capabilities of human engineering. They "learned" how to do this through a long, inefficient process of biological evolution, rather than through any process of logical reasoning, and they certainly don't know what they know, consciously.

The abilities of our machines to carry lengthy yet accurate calculations, to store massive amounts of information, and to learn by doing at an extremely fast pace are already opening up qualitatively new paths toward understanding. They will move the frontier of knowledge in directions, and arrive at places, that unaided human brains can't go. *Aided* brains, of course, can help in the exploration.

A special quality of humans, not shared by evolution or, as yet, by machines, is our ability to recognize gaps in our understanding and to take joy in the process of filling them in. It is a beautiful thing to experience the mysterious, and powerful, too.

10

COMPLEMENTARITY IS MIND-EXPANDING

> The test of a first-rate intelligence is the ability to hold two opposed ideas in the mind at the same time, and still retain the ability to function.
>
> —*F. Scott Fitzgerald*

> It is clear that this complementarity overthrows the scholastic ontology. What is truth? We pose Pilate's question not in a skeptical, anti-scientific sense, but rather in the confidence that further work on this new situation will lead to a deeper understanding of the physical and mental world.
>
> —*Arnold Sommerfeld*

Complementarity, in its most basic form, is the concept that one single thing, when considered from different perspectives, can seem to have very different or even contradictory properties. Complementarity is an attitude toward experiences and problems that I've found eye-opening and extremely helpful. It has literally changed my mind. Through it, I've become larger:

more open to imagination, and more tolerant. Now I'd like to explore with you the mind-expanding insights of complementarity, as I understand them.

The world is simple and complex, logical and weird, lawful and chaotic. Fundamental understanding does not resolve those dualities. Indeed, as we have seen, it highlights and deepens them. You can't do justice to physical reality without taking complementarity to heart.

Humans, too, are wrapped in dualities. We are tiny and enormous, ephemeral and long-lasting, knowledgeable and ignorant. You can't do justice to the human condition without taking complementarity to heart.

COMPLEMENTARITY IN SCIENCE

Niels Bohr, the great Danish quantum physicist, first articulated the unifying power of complementarity. Straightforward history would say that Bohr learned complementarity from his experience with quantum physics. A different perspective would say that this way of thinking came to Bohr naturally, predating and even enabling his unique contributions to quantum physics. Some of Bohr's biographers have seen here the influence of Søren Kierkegaard, a Danish mystic and philosopher whom Bohr admired.

Between the first inklings of quantum behavior, around 1900, and the emergence of modern quantum theory in the late 1920s, there was a period of intense struggle when it seemed impossible to reconcile different experimental observations.

During this period, Bohr was a master at building models that made sense of some observations, while strategically ignoring others. Albert Einstein wrote of his work:

> That this insecure and contradictory foundation was sufficient to enable a man of Bohr's unique instinct and tact to discover the major laws . . . of the atom together with their significance for chemistry appeared to me like a miracle—and appears to me as a miracle even today. This is the highest form of musicality in the sphere of thought.

Coming out of this experience, Bohr developed complementarity into a strong insight that flows from science into philosophy, and becomes wisdom.

COMPLEMENTARITY IN QUANTUM MECHANICS

In quantum mechanics, the most basic description of an object—whether the object is an electron or an elephant—is its wave function. An object's wave function is a kind of raw material, which we can process into predictions about the behavior of the object. We can process the wave function in different ways, in order to address different questions. If we want to predict where the object will be, we must process its wave function in one way. If we want to predict how fast the object is moving, we must do so in a different way.

These two ways of processing the wave function are broadly

similar to two ways of analyzing music, by harmony or by melody. Harmony is a local analysis—here monitoring a moment in time, rather than a point in space—while melody is a more global analysis. Harmony is like position, while melody is like velocity.

We can't do those two forms of processing at the same time. They interfere with each other. If you want to get position information, you must process the wave function in a way that destroys velocity information, and vice versa.

While the precise mathematical details can be complicated, it is vitally important to emphasize that there is a solid mathematical foundation supporting all that talk. In quantum theory, as presently understood, complementarity is a mathematical fact, not just an airy assertion.

So far, I have discussed quantum complementarity using mathematical concepts—that is, wave functions and processing. We can get a different perspective by considering the same situation more directly, in terms of experiments. In that spirit, instead of asking how we can process a particle's wave function to make predictions, we ask how we can interact with the particle to measure its properties.

Given the mathematical framework of quantum theory, the complementarity of position and velocity is a theorem. But the mathematics of quantum theory, with its many weird aspects, is an attempt to describe Nature, not revealed truth. Indeed, many of the pioneers of quantum theory, including Einstein, became skeptics of its mature mathematical form.

The counterpart of quantum theory's inability to predict position and velocity simultaneously must be our inability to measure those properties simultaneously in experiments. If it were possible to measure both position and velocity simultaneously, then we would need a new mathematical theory, different from quantum mechanics and its processed wave functions, to let us describe such measurements.

Soon after he laid the foundations of modern quantum theory, young Werner Heisenberg realized its startling mathematical consequence, that position and velocity could not be measured simultaneously. He formalized that realization as his "uncertainty principle." One of the key questions that arises from his uncertainty principle is whether or not it correctly describes concrete facts—that is, things we can observe—about the physical world. Heisenberg and then Einstein and Bohr all wrestled with this.

At the level of physical behavior, this conflict—this complementarity—reflects two key points. The first key point is that *to measure something's properties, you must interact with it.* In other words, our measurements do not capture "reality," but only sample it.

As Bohr put it:

> In quantum theory . . . the logical comprehension of hitherto unsuspected fundamental regularities . . . has demanded the recognition that no sharp separation can be made between an independent behavior of the objects and their interaction with the measuring instruments.

The second key point, heightening the first, is that *precise measurements require strong interactions.*

With those points in mind, Heisenberg considered many different ways that one might try to measure the position and velocity of elementary particles. He found, in every case, that they conformed to his uncertainty principle. That analysis built up confidence that the strange mathematics of quantum theory reflected strange facts about the physical world.

The two principles we mentioned above—that observation is an active process and that observation is invasive—were bedrock foundations of Heisenberg's analysis. Without them, we cannot use the mathematics of quantum theory to describe physical reality. They undermine, however, the world-model we build up as children, according to which there's a strict separation between an external world, which is "out there" and has properties that our observations reveal, and ourselves. Accepting the lessons of Heisenberg and Bohr, we come to realize that there is no such strict separation. By observing the world, we participate in making it.

Heisenberg did his work on uncertainty at Bohr's institute in Copenhagen. Those two pioneers had intense discussions, and developed a kind of scientific father-son relationship. Bohr's early ideas on complementarity emerged as an interpretation of Heisenberg's work.

Einstein disagreed with Bohr and Heisenberg's findings and was uncomfortable with complementarity. He was uncomfortable with the idea that there could be valid yet incompatible viewpoints. He hoped for a more complete understanding, which could encompass all possible viewpoints at once. In particular—as a test case—he hoped that both the position and the velocity of a particle could be measured simultaneously.

He thought hard about that issue, and he tried to design experiments that could reveal both the position and the velocity (or momentum*) of a particle at the same time. Einstein's ingenious thought experiments were more intricate than those Heisenberg had considered.

In the famous Bohr-Einstein debates, as described by Bohr in "Discussions with Einstein on Epistemological Problems in Atomic Physics," Einstein challenged Bohr with a series of thought experiments. These challenged aspects of quantum-mechanical complementarity, notably including the complementarity of energy and time. In responding to every one of those challenges, Bohr was able to identify subtle flaws in Einstein's analysis, and to uphold the physical consistency of quantum theory.

Those debates, and others that followed them, have clarified the nature of quantum theory, but so far they have never successfully challenged its correctness. Meanwhile, people have used quantum theory to design many wonders, from lasers to iPhones to GPS. Those quantum theory–based designs might not have worked—but they do. If "what doesn't kill you makes you stronger," then quantum theory, and the complementarity it implies, are now strong, indeed.

(In case you've been wondering what this means for the aforementioned elephant: Quantum uncertainty, although present in principle, can be safely ignored. We have no trouble

* In the preceding discussion of uncertainty, I have spoken of position versus velocity. In the physics literature, it is more common—and, for technical reasons, more convenient—to speak of momentum instead of velocity. Having inserted this footnote, I will continue to use velocity, which is more familiar to most people.

measuring both the position and the momentum of an elephant well enough to serve all practical purposes. The uncertainty in those things, compared with their actual value, is a negligible fraction. For electrons in atoms, it's a different story.)

LEVELS OF DESCRIPTION

Another source of complementarity is the use of different levels of description. When the description of a system using one kind of model gets too complicated to work with, we sometimes can find a complementary model, based on different concepts, to answer important questions.

A humble, concrete example will bring out the basic idea, which has profound implications and many applications. The gas that fills a hot-air balloon is composed of a vast number of atoms. If we wanted to predict the behavior of the gas by applying the laws of mechanics to its atoms, we'd face two big problems:

- Even if we were content to start with classical mechanics (as an approximation), we'd need to know the position and velocity of each atom at some initial time, to give the equations the data they need to work with. Gathering and storing that much data is totally impractical. Quantum mechanics only makes this problem worse.
- Even if we somehow got the data and stored it, the calculations that would be necessary to follow the particles' motions are even more impractical.

Nevertheless, practiced balloonists operate their craft with confidence. In some respects, the air behaves in easily predictable ways.

By introducing radically different concepts—density, pressure, and temperature—we can find simple laws that describe the air's large-scale behavior. It is those concepts, rather than an atomic description, which answer the questions that balloonists need answered. The atomic description contains much more information, in principle, but most of that information is worse than useless if you're interested in flying a balloon (worse, because it adds distractions). Consider, for example, the position and velocity of any particular atom. Those properties change rapidly over time, as a result of its motion and collisions with other atoms. The actual trajectory of an atom depends sensitively on the precise starting values, and also on what the other atoms are doing. Thus, information about a particular particle's position and velocity is wickedly difficult to calculate, and it goes out of date rapidly. In short, it is neither simple nor stable. Density, pressure, and temperature behave much better in those regards. It was a major scientific achievement to discover and quantify those simple, stable properties, which can be used to answer important questions.

Most of science is a search for simple, stable properties that can answer questions which interest us. We sometimes speak of these as emergent properties. (We ran into this concept before, from a slightly different angle, in chapter 7.) Finding useful emergent properties, and learning to use them skillfully, can be big achievements. The hard sciences have over their history produced many important emergent properties (entropy, chemical bond, stiffness, and so on) and built many useful models based upon them.

Similar issues arise outside the hard sciences. We'd love to have a more useful understanding of the behavior of people, or of the stock market, for example. The "atomic" versions of those subjects, working up from the behavior of individual neurons or of individual investors—let alone the behavior of the quarks, gluons, electrons, and photons that make them—are hopelessly complex. They are impractical approaches if your goal is to get along in society, or to make money by investing.

And so we turn instead to different concepts, which you will find in texts on psychology and economics, to answer our large-scale questions. They give us models of people and markets that are complementary to fine-grained, "atomic" models. In psychology and economics, we don't yet have many models that work as reliably as physicists' models of gases. The search for emergent properties, and for useful models built up from them, continues.

There is immense satisfaction in describing the world in terms of its most elementary building blocks. It is tempting to say that this is the ideal description, while other, high-level descriptions are mere approximations—compromises, which reflect weakness in understanding. That attitude, which makes the perfect the enemy of the good, is superficially deep, but deeply superficial.

In order to answer questions of interest, we often need to change focus. To discover—or invent—new concepts, and new ways of working with them, is an open-ended, creative activity.

Computer scientists and software engineers are well aware that in designing useful algorithms, it is important to pay attention to how knowledge is represented. A good representation can make the difference between usable knowledge and knowledge that is there "in principle," but not really available, because it takes too long and too much trouble to locate and process. It's like the difference between owning bars of gold and knowing that in principle there are vast stores of gold atoms floating dissolved in the ocean.

For that reason, complete understanding of the fundamental laws, if we ever achieved it, would be neither "the Theory of Everything" nor "the End of Science."* We would still need complementary descriptions of reality. There would still be plenty of great questions left unanswered, and plenty of great scientific work left to do.

There always will be.

BEYOND SCIENCE: COMPLEMENTARITY AS WISDOM

Examples from Art

My musical friend Minna Pöllänen brought up a beautiful example of complementarity in her domain, which I briefly mentioned earlier. In polyphonic music, two very different things occur together—each voice carries a tune, while the ensemble

* Those are two phrases, endemic in popular science journalism, that I find extremely irritating.

moves through harmonies. We can focus on the melodies or focus on the harmonies. Each is a meaningful way to interact with the music. You can switch between them. But you can't really do both at once.

Picasso and the Cubists created visual art that captures complementarity pictorially. By taking up different perspectives on a scene in the same picture, they were liberated to bring out with great freedom aspects they feel are important. Young children do this, too, in their drawings. The bizarre exaggerations and juxtapositions in these artworks emphasize different views that could be considered contradictory. In the physical world, they could not be realized simultaneously. Such up-front complementarity can be charming in a child's drawings, and genius in a master's.

Models of People—Free and *Determined*

We construct mental models of people, too, as ways of answering questions about them. For example, if we want to predict how someone will behave in a social situation, we might consider their personality, their emotional state, their life history, the culture they were born into, and so forth. In short, we construct a model of their mind and motives. The concept of will—a mind making choices—is central to this model.

On the other hand, if we want to predict what will happen to that same person if they are at ground zero of a nuclear explosion, then quite a different model, based on physics, will be appropriate. In that case, mind and will don't come into it at all.

Both models—one based on mind and psychology, the other based on matter and physics—are valid. Each addresses a different question successfully. But neither is complete, and neither makes a good substitute for the other. People *do* make choices, and their bodies *are* subject to the rules of matter. Those observations are everyday facts. They won't go away. In the spirit of complementarity, we accept them both. We recognize that neither falsifies the other. Facts can't falsify other facts. Rather, they reflect different ways of processing reality.

Do people have choice in what they do, or are they puppets who dance to the tune of mathematical physics? That is a bad question, not unlike asking whether music is harmony or melody.

Free will is an essential concept in law and morality, while physics has been successful without it. Removing free will from law, or injecting it into physics, would make a mess of those subjects. It is totally unnecessary! Free will and physical determinism are complementary aspects of reality.

Complementarity, Mind Expansion, and Tolerance

Let me re-express, in simpler terms, the basic messages of complementarity:

- The questions you want answered mold the concepts you should use.
- Different, even incompatible, ways of analyzing the same thing can each offer valid insights.

Thus, complementarity is an invitation to consider different perspectives. Unfamiliar questions, unfamiliar facts, or unfamiliar attitudes, in the spirit of complementarity, give us opportunities to try out new points of view and to learn from what they reveal. They foster mind expansion.

Why not bring this spirit to supposed conflicts between art and science, or philosophy and science, or religion A and religion B, or religion and science?

It can be illuminating to look at the world in different ways.

In my own experience, early exposure to Catholicism inspired me to think cosmically and to look for hidden meanings beneath the appearance of things. Those attitudes have proved enduring blessings, even after I abandoned the faith's strict dogmas. Today, I often go back to Plato, to Saint Augustine, to David Hume, or to "outdated" original scientific works—Galileo, Newton, Darwin, Maxwell—to converse with great minds, and to practice thinking differently.

Of course, trying to understand different ways of thinking does not necessarily mean you must agree with them, much less adopt them as your own. In the spirit of complementarity, we should maintain detachment. Ideologies or religions that claim an exclusive right to dictate uniquely "correct" views are contrary to the spirit of complementarity.

That said, science has a special status. It has earned enormous credibility, both as a body of understanding and as an approach to analyzing physical reality, through its impressive success in many applications. Scientists who define themselves narrowly fail to enrich their minds, but people who avoid science impoverish theirs.

THE FUTURE OF COMPLEMENTARITY

Accuracy and Comprehensibility

The rise of supercomputers and artificial intelligence is changing both the kinds of questions we can ask and the kinds of answers we can seek.

Bohr himself referred, half-jokingly, to the complementarity between clarity and truth. This goes too far, since there are certainly things, like the basics of arithmetic, that are both clear and true.

But successful models that require superhuman computations open up an analogous complementarity, which is quite serious. In chess and Go, two games whose mastery was once thought to represent the pinnacle of intelligence, computers are now the best players.

Each of those games has a large literature, wherein great human players explain the concepts they've used to organize their knowledge. The present-day champions—computers—don't use those concepts. The human concepts are adapted to brains with tremendous powers to use imagery and do parallel processing, but that have relatively weak memories and run at relatively sluggish speeds. A computer can develop entirely different concepts, and also discover the effective human concepts, simply by playing games against itself many, many times and observing what works—in other words, by following the scientific method of learning from experiments.

In quantum chromodynamics, our theory of the strong interaction, people invented concepts to bridge the gap between the basic equations for quarks and gluons and the more complex objects that finally appear in Nature. Those concepts have helped human minds to get a grip on the problem. To date, however, the strategy that's worked best—by far—is to hand over the calculations, with minimal instructions, to supercomputers.

Those examples are distinguished by their clarity (and truth), but the basic phenomenon they exemplify, that thinking machines can discover and use models that are impractical for unassisted human brains, is likely to be widespread.

In short: Human comprehensibility and accurate understanding are complementary.

Humility and Self-Respect

The complementarity between humility and self-respect is, I believe, the central message of our fundamentals. It recurs as a theme in many variations. The vastness of space dwarfs us, but we contain multitudes of neurons, and, of course, vastly more of the atoms that make up neurons. The span of cosmic history far exceeds a human lifetime, but we have time for immense numbers of thoughts. Cosmic energies transcend what a human commands, but we have ample power to sculpt our local environment and to participate actively in life among other humans. The world is complex beyond our ability to grasp, and rich in mysteries, but we know a lot, and are learning more. Humility is in order, but so is self-respect.

Many decades may pass before autonomous, general-purpose artificial intelligences (AIs) reach human levels. But so powerful are the motivations, and so inexorable is the progress, that barring catastrophic wars, climate change, or plagues, it is likely to take only a century or two. Given the intrinsic advantages in speed of thought, strength of perception, and physical power that engineered devices can offer, the vanguard of intelligence will pass from lightly adorned *Homo sapiens* to cyborgs and super-minds.

It is also possible that genetic engineering will produce creatures of superhuman abilities. They will be smarter, stronger, and (I hope and expect) more empathetic than present-day humans.

To realize that these looming possibilities do, in fact, loom adds, for thinking humans today, a new dimension of humility. Yet self-respect is still in order. In a moving passage from his 1935 novel *Odd John*, science fiction's singular genius Olaf Stapledon has his hero, a superhuman (mutant) intelligence, describe *Homo sapiens* as "the archaeopteryx of the spirit."* He says this fondly to his friend and biographer, who is a normal human.

Archaeopteryx was a noble creature and, I suspect, not an unhappy one. Flying—perhaps badly, but better than your fellow creatures, and better than your ancestors—is a heady experience. The glory of archaeopteryx is enhanced, not diminished, by the brilliance of its descendants.

---

* Archaeopteryx was a species with both dinosaur-like and bird-like features, linking dinosaurs that were bound to the earth and the birds we admire in the air today.

AFTERWORD: THE LONG VOYAGE HOME

The fundamentals of science are not comfortable. As they teach us, they challenge our habits of thought. Most profoundly, they raise the bar for what we should expect from true understanding. They raise it so high as to make the understanding we have achieved seem eternally inadequate. This is the meaning of John R. Pierce's ironic observation that "we will never again understand nature as well as Greek philosophers did."

The fundamentals of science can undermine faith in received beliefs and conventional wisdom. In particular, they make it difficult to take mythological stories about natural phenomena seriously. It has become all but impossible to believe that Apollo pulls the Sun across the sky with his chariot.

That undermining process can go much further, beyond merely discrediting absurdities. Scientific understanding bears such abundant and delightful fruit that eating from its Tree of Knowledge can spoil one's taste for other foods. Nonscientific literature can come to seem stale; nonscientific philosophy silly; nonscientific art pointless; nonscientific traditions hollow—and, of course, nonscientific religion nonsensical. During my early teenage years, in my first heady engagement with modern science, those were my attitudes.

If a painful narrowing of one's outlook was the price of accepting the scientific fundamentals, many people would reasonably conclude that the price is too high. Thankfully, the fundamentals of science do not require you to make those corrosive applications of science.

Science tells us many important things about how things are, but it does not pronounce how things should be, nor forbid us from imagining things that are not. Science contains beautiful ideas, but it does not exhaust beauty. It offers a uniquely fruitful way to understand the physical world, but it is not a complete guide to life.

On calmer reflection, I began to appreciate those facts. Over time, I've come to feel their truth ever more deeply.

The child of our introduction, now an adult, may come to understand the fundamental conclusions that science, following its radically conservative method, reaches about the physical world. Then she is prepared to revisit the starting point of her adventure with reality, and to view it afresh, in the light of her knowledge. She can choose, in this sense, to be born again.

It is not a trouble-free choice. It is disruptive. But the choice is unavoidable, as a matter of integrity. You've seen in this book a small sampling of the evidence for the scientific fundamentals. That evidence is overwhelming and indisputable. To deny it is dishonest. To ignore it is foolish.

And so our heroine comes to reconsider the division of experience into internal and external worlds. The fundamentals of science have taught her a lot about what matter is. She knows that matter is built up from a few kinds of building blocks, whose properties and behavior we understand in detail. And she knows, from direct experience, that scientists and engineers can use such knowledge to make impressive creations. Her iPhone allows her to communicate instantly with friends around the globe, to tap into humanity's accumulated knowledge at will, and, through pictures and recordings, to snatch her sensory world from time's devouring flow.

She has learned, too, that the special objects she recognizes as other people, and herself, are made from the same sort of matter as the rest of the world. Many once-mysterious aspects of living things, such as how they derive their energy (metabolism), how they reproduce (heredity), and how they sense their environment (perception), she can now understand from the bottom up. For we now understand, in considerable detail, how molecules—and ultimately, quarks, gluons, electrons, and photons—manage to accomplish those feats. They are complicated things that matter can do, by following the laws of physics. No more, and no less.

These understandings do not subtract from the glory of life. Rather, they magnify the glory of matter.

In light of all this, it is radically conservative to adopt what the great biologist Francis Crick has called "the astonishing hypothesis": that mind, in all its aspects, is "no more than the behavior of a vast assembly of nerve cells and their associated molecules." Indeed, this amounts to extending Newton's method of analysis and synthesis to brains. Experimenters in neurobiology have been following that strategy aggressively. And although our understanding of how minds work is still incomplete, so far, in thousands of sensitive experiments, the strategy has never failed. No one has ever stumbled upon a power of mind in biological organisms that is separate from conventional physical events in their bodies and brains. Even in their most delicate experiments, physicists and biologists never had to make allowances for what people nearby were thinking. By now, any failure of Crick's "astonishing hypothesis" would be astonishing.

Upon that realization, the division of experience into internal and external worlds comes to seem superficial. For babies, that division is a useful discovery, and for adults, it is a convenient rule of thumb. But our best understanding suggests that there is just one world, after all. Matter, deeply understood, has ample room for minds. And so, also, it can be home to the internal worlds that minds house.

There is both majestic simplicity and strange beauty in this unified view of the world. Within it, we must consider ourselves not as unique objects ("souls"), outside of the physical world, but rather as coherent, dynamic patterns in matter. It is an unfamiliar perspective. Were it not so strongly supported by the fundamentals of science, it would seem far-fetched. But it has the virtue of truth. And once embraced, it can come to seem liberating. Albert Einstein spoke to this, in a kind of credo:

> A human being is part of a whole, called the Universe, a part limited in time and space. He experiences himself, his thoughts and feelings, as something separated from the rest, a kind of optical delusion of his consciousness. This delusion is a kind of prison for us.

I have been at pains to be clear that science teaches us what is, not what ought to be. Science can help us attain our goals, once they are chosen, but it does not choose our goals for us.

Still, in this last section, I'd like to make a connection between the unified view of the world our heroine has achieved and a moral attitude. The connection will not be a scientific proof. What recommends it is its harmony.

Notoriously, views of morality have changed over time. (Here I am looking backward, from the perspective of American culture in the early twenty-first century.) Based on experience and consensus, people have gradually abandoned old views and adopted new ones. Thus, it is fair to say that, judged by experience and consensus, the new views are improvements on the old ones. Slavery was taken for granted by many in the ancient world, but now it is almost universally condemned, as are racism, sexism, nationalistic aggression, and cruelty to animals. A common theme in all these developments is a widening circle of empathy. With progress, we've come to consider people and creatures as having intrinsic value and being worthy of profound respect, just like ourselves. When we see ourselves as patterns in matter, it is natural to draw our circle of kinship very wide, indeed.

Here is the continuation of Einstein's credo:

> [This delusion is a kind of prison for us], restricting us to our personal desires and to affection for a few persons nearest us. Our task must be to free ourselves from this prison by widening our circles of compassion to embrace all living creatures and the whole of nature in its beauty.

Those tasks of liberation and empathy are not separated from understanding the fundamentals of science. Indeed, understanding helps us to achieve them. The universe is a strange place, and we're all in it together.

Acknowledgments

I have been blessed throughout my life with wonderfully supportive parents, family, teachers, and friends too numerous to mention individually. It seems appropriate, though, to especially acknowledge my debt to the public school system of New York City.

Alfred Shapere, Wu Biao, Thomas Houlon, and Patty Barnes read this book in draft form and gave valuable feedback. I worked closely with Christopher Richards and Elizabeth Furlong as editors, and got help from many others at Penguin Press. John Brockman, Katinka Matson, and Max Brockman encouraged me in this project and helped me to see it through.

Appendix

In this appendix, I've gathered together brief discussions of some informative material that supplements the main text but that seemed either tangential to the discussion or too technical for the spirit of this book.

MASS AS A PROPERTY

Mass plays a role in two aspects of a particle's behavior, governing both its inertia and its gravity. The inertia of a body measures its resistance to changes in its motion. Thus, a body that has large inertia will tend to keep moving at its present velocity unless it is subjected to large forces. The gravity of a particle is a universal attraction it exerts on other particles. The larger the mass of a particle, the larger its gravity. Each kind of elementary particle has a definite value for its mass. The values for different particles are generally different. They don't appear to fit into any simple pattern. Many physicists have tried to explain the observed values of elementary particle masses, but nobody has succeeded.*

Some of the most important particles, including photons, gluons, and gravitons, have zero mass. This does not mean that they have no inertia, or that they exert no gravity. In fact, they do. Let me explain that paradox, which in my experience often troubles thoughtful learners.

Mass contributes to inertia and gravity, but it is not the only factor. In particular, a moving particle has more inertia, and exerts more gravity, than a particle at rest. Indeed, the theory of relativity teaches us that it is energy, not mass, that controls inertia and gravity. For bodies at rest, energy and mass are proportional, according to Einstein's famous formula $E = mc^2$, so in that case we can express inertia and gravity using either one, interchangeably. When bodies move slowly, relative to the speed of light, $E = mc^2$ remains true to a good approximation. In that case, we don't make a big mistake if we say that inertia and gravity are proportional to mass.

For bodies whose speed is close to the speed of light, however, $E = mc^2$ is way off. It's not that Einstein blundered, but that a more general and sophisticated version of the formula, also devised by Einstein, should be used. The more general formula shows that photons carry energy, and thus that they have inertia and exert gravity, despite having zero mass.

* More precisely, nobody has succeeded in convincing anybody else that they've succeeded.

CHARGE AS A PROPERTY

A particle's electric charge governs the strength with which it participates in the electromagnetic force. We've explored the nature of that force in the main text. Here we focus on electric charge itself, as a property of elementary particles.

Two facts about electric charge make it especially easy and pleasant to work with. One is that it is additive—which is to say that you can calculate the total electric charge of a collection of objects simply by adding up the electric charges of its component parts. The second is that it is conserved. This means that the total electric charge in an isolated region of space will stay the same no matter what happens within that region. The charge can change if you bring things in or take them out, but not if you rearrange them or bash them into one another.

Quantities that are additive and conserved embody the intuitive notion of "substance." They add up and don't get lost. You can literally count on them.

The electric charges of elementary particles follow a much simpler and more regular pattern than do their masses. Many elementary particles have zero electric charge, and all the nonzero charges are whole-number multiples of a common unit.* Some are positive, and some are negative.

A body's electric charge, as I mentioned, governs the strength of its response to electric and magnetic fields. There are two other kinds of charge, analogous in many ways to electric

* This is a third pleasant property of electric charge. Physicists say, somewhat confusingly, that it is "quantized."

charge, that play a similar role in the other fundamental interactions. They are called color charge and weak charge.

A body's color charge governs the strength of its response to gluon fields. I like to say that color charge is like electric charge, but on steroids. The unit of color charge, which governs the strength of the strong force, is bigger than the unit of electric charge (that is, the charge of the electron). This is what makes the strong force strong. Not only that, but there are three different kinds of color charge, and eight different kinds of gluons that respond to them, as opposed to one kind of electric charge and one photon.

Altogether, the system of equations that govern the strong force, known as quantum chromodynamics (QCD), is a larger, more symmetrical version of Maxwell's equations, which govern quantum electrodynamics (QED), the modern theory of electromagnetism. QCD is QED on steroids.

Weak charge comes in two kinds, and their unit is slightly larger than the unit of electric charge. The physical significance of weak charge becomes clear only within the context of ideas around the Higgs condensate, as featured in chapter 8.

PARTICLES OF CHANGE

What I've called the particles of change are of two sorts. *W* and *Z* bosons, and the Higgs boson, are about a hundred times heavier than protons. They are also highly unstable. These two facts—their heaviness and their instability—imply that they are both difficult to produce and transient. Their production and detection was a major achievement of work at high-energy accelerators in recent decades. Neutrinos are very light and they are basically stable, but they interact very feebly with ordinary matter (that is, matter made from the particles of construction). Here is a table, parallel to the similar one for particles of construction in the main text:

	mass	electric charge	color charge	spin
neutrinos (3 kinds)	< .00001	0	no	½
W	157,000	1	no	1
Z	178,000	0	no	1
Higgs particle	245,000	0	no	0

Though they are not significant ingredients of ordinary matter, these particles play a crucial role in the natural world. They are involved in processes of transformation: the so-called weak interaction, or weak force. In the natural world, energy released in some of these weak force processes drives plate tectonics and gives stars their power. It also makes nuclear reactors and nuclear weapons possible.

There are three kinds of neutrinos, distinguished by different masses and subtly different interactions. They are all extremely light. As indicated in the table, their masses are a tiny fraction of the electron's, but in at least two cases (and probably all three) it is not zero. Since they have zero electric charge and no color charge, neutrinos interact feebly with ordinary matter. This makes them difficult to study. When

Wolfgang Pauli proposed, for theoretical reasons, the existence of neutrinos, he didn't write a regular journal article about it. Instead, he sent a jocular letter to a conference of nuclear physicists. To a friend, he expressed this self-reproach: "I have done something very bad today by proposing a particle that cannot be detected; it is something no theorist should ever do."

But experimenters rose to Pauli's backhanded challenge by building and instrumenting gigantic detectors. Today, neutrino physics is a thriving experimental activity. It gives us, among other things, clear looks into the Sun's core and into the violent transformations that power supernova explosions.

Finally, the Higgs particle is described at length in chapter 8, where it is a featured player.

BONUS PARTICLES

Now we come to a group of elementary particles nobody really knows what to make of. The bonus particles are all unstable. They were discovered among the debris of high-energy collisions, either in cosmic rays (early in the twentieth century) or at particle accelerators (more recently). When the first of them, the muon, was discovered in 1936, the renowned physicist I. I. Rabi captured the community's bewilderment in a quip that's become legendary: "Who ordered that?"

The masses of these bonus particles span a wide range and form no obvious pattern, as you can see from the following table.

	mass	electric charge	color charge	spin
c quark	2,495	⅔	yes	½
t quark	339,000	⅔	yes	½
s quark	180	-⅓	yes	½
b quark	8,180	-⅓	yes	½
muon	207	-1	no	½
tauon	3,478	-1	no	½

These particles form three groups. Looking at their properties, you'll see that the *c* and *t* quarks are heavier, unstable versions of the *u* quark, while the *s* and *b* quarks are heavier, unstable versions of the *d* quark, and the muon and tauon are heavier, unstable versions of the electron.

Our final "elementary particle" is a work in progress. Astronomers have observed, in many situations, more gravity than they can account for. It is not a small discrepancy: To get the observed gravity, we need about six times more mass than ordinary matter provides. This is the so-called dark matter problem, as described in chapter 9.

An elementary particle with the right properties could solve the dark matter problem, by providing a source for the otherwise mysterious gravity. The observed facts are broadly consistent with that explanation, but they don't provide enough information to pin down crucial properties of the particle, such as its mass and spin.

	mass	electric charge	color charge	spin
dark matter	unknown	0	no	unknown

FOR MORE INFORMATION: A GO-TO CATHEDRAL

The website of the Particle Data Group is http://pdg.lbl.gov. It chronicles and documents the empirical evidence for our fundamental understanding of cosmology and of matter and its interactions in full technical detail. It is a scientific cathedral, dutifully erected by a human community spanning several generations and all of Earth's continents, in tribute to the glory of physical reality.

QCD LAID BARE: JETS

The strong force among quarks and gluons becomes feeble not only for *small* separations in time and distance, but also for *large* changes in energy and momentum. These behaviors are two facets of asymptotic freedom. Using the equations of quantum mechanics, either one can be derived from the other.

The rarity of large changes in energy and momentum leads us to a striking phenomenon, which has emerged as a dominant feature of ultra-high-energy interactions. This is the phenomenon of *jets*. Jets lay bare the essence of QCD. They exhibit quarks, gluons, and their basic interactions in an amazingly direct, tangible form.

Let's consider what happens when a quark within a proton is suddenly jerked by an external force. The external force might come from a bombarding electron, for example. The quark, ripped from its normal environment, starts with a lot of energy and momentum, and leaves the proton. An isolated quark is an untenable situation, however. Its uncompensated color charge interferes with the equilibrium of the color gluon fields, and the quark thereby radiates gluons, shedding energy and momentum. Those secondary gluons will also radiate, emitting either gluons or quarks and antiquarks. In this way, the initial jerk leaves a trail of quarks, antiquarks, and gluons, which then congeal into protons, neutrons, and other hadrons. As always, the quarks, antiquarks, and gluons do not materialize as individual particles, but only within associations (hadrons).

This might sound like complicated business, and it is. But asymptotic freedom gives structure to the mess. Since radiation that involves large transfers of energy and momentum is rare—that's what asymptotic freedom says—all the particles in the cascade tend to be moving in the same direction. In the end, we observe many particle tracks emerging within a narrow cone. We say they make a *jet*. Since energy and momentum are conserved, overall, the total energy and momentum of all the particles within our jet add up to the energy and momentum of the original quark.

Jets are a wonderful gift to physicists. Because they encode the energy and momentum of the particles that initiated them, they serve as avatars for those particles. In this way, quarks and gluons become quite tangible things, even though they themselves do not exist as isolated particles. We can translate predictions for quark and gluon behavior into predictions for jets. Jets thereby allow us to check the basic laws of QCD, which are statements about quarks and gluons, precisely and in great detail. They also give us a handle on other processes, known or hypothetical, that involve quarks and gluons.

It is standard practice for experimenters to report how many quarks and gluons are produced in the reactions they study, how they are distributed in energy and angle, and so forth. What they've actually observed is the corresponding jets, but the identification, after thousands of successful applications, has become routine. Quarks and gluons entered the world as weird, suspect theoretical phantoms—confined particles that, according to theory, would never be observed in isolation. Tamed by beautiful ideas, they've become tangible realities—not mere particles, but jets.

GEOMETRY OF SPACE AND DENSITY OF MATTER

General relativity predicts a striking relationship between the average curvature of space, the average density of matter within it, and the rate of expansion of the universe. If the total density of matter is equal to a certain critical density, then space will be flat; if the density is larger, it will be positively curved, like a sphere; if the density is smaller, it will be negatively curved, like a saddle.

At present, the critical density is about 10^{-29} grams per cubic centimeter. This is equivalent to the mass of about six hydrogen atoms per cubic meter. Though this critical density is far below the density of the best "ultra-high vacuum" people have achieved in laboratories on Earth, it seems that it is close to the average density of the universe as a whole.

Astronomers can measure the shape of space geometrically, using sophisticated versions of the procedures we indicated in chapter 1. They can also measure the density, by adding up contributions from ordinary matter, dark matter, and dark energy. They find that space is very nearly flat, and that the density is very nearly the critical density. This is consistent with the prediction of general relativity. That consistency encourages us to think that the dark matter and dark energy mysteries can be understood within the framework of general relativity. Certainly, they do not *require* its modification.

Index